# Raising Future Innovators
Leveraging Jewish & Chinese Best Practices in Education

Ami Dror & Jordan Huang

Copyright © 2019
by Ami Dror & Jordan Huang

All rights reserved. No part of this book may be reproduced or used in any manner without written permission of the copyright owner except for the use of quotations in a book review. For more information, please contact us at: info@raisingfutureinnovators.com

FIRST EDITION

www.raisingfutureinnovators.com

# Preface

As we face a dynamic, fast-paced and technology-driven future, it is now more apparent than ever that innovation and entrepreneurship will be critical to our children's success. From global health to transnational migration, climate change to economic stability, innovation lies at the center of our ability to achieve progress as individuals, communities and nations. This book was written with the goal of providing parents and educators around the world with important insights into some of the best practices in education that can help cultivate these essential 21$^{st}$ century skills. By leveraging the wisdom of both Chinese and Jewish educational perspectives, we hope to empower parents on their quest to raise the next generation of leaders. Together, we believe that we can create the educational experiences and foundation for skill-acquisition that will increase innovation across the globe.

While our focus within this book stems from examining elements of the Chinese education system, we believe that our unique perspective will be beneficial to educators across all cultures. International testing data shows that American high schoolers perform at a distinctly mediocre level in reading, math, and science, particularly in comparison with Chinese students who scored at the very top. Beyond achieving impressive test scores, over the past decade, the Chinese government has begun taken forward-thinking steps towards educating students for the future. Both Chinese and Western thought-leaders are coming to terms with the fact that in order for students to compete on the world stage, they will need more than excellent test scores. They must be able to adapt and problem-solve through new, complex challenges – **they will need to be innovators.**

While this book references an array of academic research and renowned sources in order to provide examples and background for the ideas presented, this work is largely the brainchild of our respective, multi-faceted experiences as Jewish and Chinese individuals. The concepts and recommendations provided are the result of our cumulative decades of personal and professional experience in the fields of education, entrepreneurship and innovation. We acknowledge the finite scope and the natural limitations of our work, while also recognizing and upholding the value of our unique perspective. We believe that parents and educators will find great value in the educational methodologies and "best practices" put forth throughout the book.

# Acknowledgements

We would like to thank both of our families for their support throughout the process of writing this book and all of our other endeavors. We would like to give special thanks to each of our mothers, without whom we wouldn't be who we are. The commonalities we share as authors, entrepreneurs and citizens of the world have their roots in the experiences we have had growing up with strong, determined mothers. Both our own mothers and the mothers of future innovators we hope to reach with this book have served as inspiration throughout our writing process.

To our many colleagues who have provided us with their invaluable feedback, we thank you for your wisdom and encouragement. We are blessed to be part of unique professional communities that have played an instrumental role in our lives. We'd like to thank all of our friends from the Aspen Global Leadership Network and the Alumni Associations of our alma maters. Our optimism and faith about the future of innovation stems, in no small part, from seeing the work of our colleagues from these global networks.

Our professional teams at LeapLearner and Uni-Ed have of course been instrumental in the creation of this book. We'd like to thank our many co-workers for enabling to learn and grow from the cross-cultural experiences we've had within our roles as founders and CEOs of these companies. These experiences inspired us to find a way for us to translate our knowledge into a more innovation-focused education for students in China and around the world.

Lastly, we'd like to thank our editor, Lerone Lessner, for her long-term commitment to this project and hard work in helping us translate our knowledge, ideas, memories and experiences into a fully realized book.

# Table of Contents

| | |
|---|---|
| Introduction | 9 |
| Chapter 1: An Exploration of Common Values | 17 |
| Chapter 2 : What is Innovation? | 25 |
| Chapter 3: From Survival to Success – | |
| The History of Modern European and American Jewry | 39 |
| Chapter 4: Passion, Creativity & Innovation in Early Education | 45 |
| Chapter 5: The Refugee | 51 |
| Chapter 6: Language | 61 |
| Chapter 7: "Balagan" - Chaos | 71 |
| Chapter 8: Debate | 81 |
| Chapter 9: "Chutzpah" – Heedless Courage and Insight | 95 |
| Chapter 10: Risk Taking | 107 |
| Chapter 11: Military Service | 119 |
| Chapter 12: Travel, Learning & Destiny | 129 |
| Chapter 13: Distinct Traits of Jewish Nobel Prize Winners | 145 |
| Chapter 14: Conclusions | 163 |
| Author Biographies | 169 |
| References | 171 |

# Introduction

As a cultural group, Jews make up approximately 0.2% of the world's population. Despite this extremely small percentage, it is impossible to ignore the contributions of Jewish individuals to our global society. Perhaps the most prominent illustration of such contributions is the number of Jewish Nobel Laureates. Since the establishment of the Nobel Prize in 1895, approximately 22% of individual winners have people of Jewish origin. Since the turn of the century, Jews have been awarded 26% of all Nobel Prizes and 28% of those in the scientific research fields. By contrast, China has a population of 1.3 billion people that comprises over 18% of the world's population and yet, to date, only 1% of Nobel Prize winners have been of Chinese descent.

The presence of Jewish excellence across all subjects and sectors has been the subject of thousands of books, journals and academic research. Everyone from psychologists and scientists to educators and business executives have developed theories to explain the success, leadership, and innovation cultivated by those of Jewish origin. Similarly, hundreds of books have been written about the Jewish experience and how it has been shaped by ancient Jewish values, education, and the complex history of the Jewish people. With a similar goal of understanding how history, education and values can breed a culture of achievement, in recent years the world has turned much of its attention to China's critical role in our global society and the unparalleled growth of the Chinese economy over the last few decades. Given China's historic timeline and international leadership within our global market, recent studies have gone to great lengths to analyze and study China's internal accomplishments as well as its success in world markets.

This book sets out to explore the overlap and connection between these two unprecedented, important global phenomena. What follows is an in-depth look at the ways in which the paths, values and history of the Jewish and the Chinese peoples have converged and diverged throughout the ages. We will offer an examination of how these two ancient civilizations innovate and create. Over the next 13 chapters we will shed new light on aspects of Jewish culture and identity that contributed to their success as global innovators. We will outline what we believe to be the pillars of success among Jewish Nobel winners and Jewish scholars throughout history. We seek to take existing analyses one step further in order to show how China might learn from the innovation of the Jewish people in order to cultivate the next generation of creative thinkers and Chinese Nobel Prize Winners. Additionally, we will compare and contrast the modern Jewish student and the modern Chinese student with the goal of highlighting how each can take steps towards achieving true innovation and receiving world renowned awards such as the Nobel Prize.

By analyzing the similarities and differences between Jewish and Chinese cultures, perspectives and approaches to education, we can translate the unique experience of the Jewish people and their innovative spirit into China's educational system. Through the identification of key factors that contributed to Jewish experiences, values, and system of education we can also enable Chinese educators to more effectively cultivate innovation among their students.

# The Nobel Prize

As we begin the exploration of these ideas, we turn first to a brief history of the roots of the Nobel Prize and its awardees over the last approximately 120 years. The prize was established by Alfred Nobel in 1895. Nobel was born in Stockholm, Sweden in 1833 to a family of engineers and grew up studying a range of subjects including chemistry and engineering. He was interested in inventions that would assist armaments manufacturers in the modernization of warfare and would become most well-known for the invention of dynamite. However, it is perhaps what he bequeathed to mankind in his will with the establishment of the Nobel Prize that stands out most in his biography.

As he neared his death, Nobel looked back on his life of armaments development and became well aware of what his legacy would look like. He therefore fashioned a last will and testament designed to promote peace and mutual understanding within a world that he felt had become filled with inhumanity and war. In an excerpt from his will, Nobel writes that the remaining funds from his estate should be used to endow "prizes to those who, during the preceding year, shall have conferred the greatest benefit to mankind." The full text regarding the establishment of the Nobel prize reads as follows:

*The whole of my remaining realizable estate shall be dealt with in the following way: the capital, invested in safe securities by my executors, shall constitute a fund, the interest on which shall be annually distributed in the form of prizes to those who, during the preceding year, shall have conferred the greatest benefit to mankind. The said interest shall be divided into five equal parts, which shall be apportioned as follows: one part to the person who shall have made the most important discovery or invention within the field of physics; one part to the person who shall have made the most important chemical discovery or improvement; one part to the person who shall have made the most important discovery within the domain of physiology or medicine; one part to the person who shall have produced in the field of literature the most outstanding work in an ideal direction; and one part to the person who shall have done the most or the best work for fraternity between nations, for the abolition or reduction of standing armies and for the holding and promotion of peace congresses. The prizes for physics and chemistry shall be awarded by the Swedish Academy of Sciences; that for physiology or medical works by the Karolinska Institute in Stockholm; that for literature by the Academy in Stockholm, and that for champions of peace by a committee of five persons to be elected by the Norwegian Storting. It is my express wish that in awarding the prizes no consideration be given to the nationality of the candidates, but that the most worthy shall receive the prize, whether he be Scandinavian or not.*

> **Key Facts about the Nobel Prize**
> - 896 Laureates and 27 organizations have been awarded the Nobel Prize between 1901 and 2017.
> - The Nobel Prize amount for 2017 is set at Swedish kronor (SEK) 9.0 million per full Nobel Prize., or approximately 7.05 Million RMB and approximately US $1.07 Million.

The Nobel Prize is widely considered a standard for recognizing innovation and excellence. It is perhaps the world's most accepted definition of success in any particular field. In examining the lives, careers and educational background of successful and innovative Jewish Nobel Laureates to date, we seek to offer a new interpretation of what the path to success might look like for today's Chinese students. One of the early successful Jewish Nobel Laureates that sets the example and tone for this research is Jewish Nobel laureate Isidor Isaac Rabi. Born in 1898 in what is now modern-day Poland, he moved with his parents to the USA when he was only one year old. Among many other accomplishments, Rabi was awarded the Nobel Prize for Physics in 1944. In an interview in *Parents Magazine* in September 1993, Rabi discussed his greatest educational influences:

*"My mother made me a scientist without ever intending to. Every other Jewish mother in Brooklyn would ask her child after school: 'So? Did you learn anything today?' But not my mother. 'Izzy,' she would say, 'did you ask a good question today?' That difference — asking good questions — made me become a scientist."*

Rabi believed that great minds start with questions. He also believed that his success is owed, at least in part, by the values his parents instilled in him at an early age. In the coming pages we seek to illuminate **what it means to ask a "good question."** By clearly defining the core tenants of such questions, we can enable parents and teachers to use them to help students become innovative thinkers. We believe that by cultivating a culture of asking "good questions," we can instill in young thinkers a sense of wonder and purpose. In our analysis of the paths taken by many Jewish students, scholars, and start-up CEOs, we will explore what role the concept of "good questions" may have played on their paths to success.

As educators and parents across the world grapple with the challenge of ensuring children acquire the skills to remain competitive in the 21$^{st}$ century, we recognize the complexities inherent to educating our future innovators. This issue has gained particular significance within the Chinese educational system where students face many hurdles as they strive to achieve top marks and succeed academically. We acknowledge and understand that China is facing a fundamental challenge in the quest to successfully educate its youth. Parents and educators are now asking themselves: How can we balance the emphasis on creating value and pride among family, party and country, with the desire to let children reach their potential and discover the strengths that make them unique? Chinese students today are pursuing the "Return to the China Dream" and are striving to be the best in the world. This is of course a noble cause. However, to lead a fulfilling life of self-enrichment, we must find a way to integrate the Chinese goals of bringing glory to the family, party and country, with a path similar to that of many well-renowned Jewish innovators – one fueled by passion, self-motivation and grit.

While we seek to provide insight and practical tools for Chinese parents and educators alike, the goal and focus of this book is not to offer commentary about the current structure of schooling in China today. Likewise, there is no proposed "winning recipe for success" in navigating the current Chinese educational system. Instead, this book focuses on developing a set of innovation-oriented, educational tools and insights. In leveraging their vastly different lives, diverse educational experiences and notable business acumen, the authors have been able to put forth ideas and recommendations that have clear international application. However, this book was conceived with the understanding of the critical role China plays in our global society and as such strives to provide unparalleled value for the Chinese parents and educators of future Chinese Nobel Laureates.

**Who We Are**

This book was co-written by **Ami Dror and Dr. Zhaodan (Jordan) Huang.** Written from the perspective of two successful entrepreneurs – one Israeli, one Chinese – this book enables its readers to form a deep understanding of Jewish culture as it relates to the cultivation of innovation, creativity and disruptive problem-solving. Coming together from very different backgrounds, this work seeks to present a truly unique collaboration that highlights "the best of both worlds" for the reader.

**Dror**, born in the south of Israel, is a self-taught computer coder and serial entrepreneur. He discovered technology as a young child when his mother bought him a simple textbook about computers. He soon convinced his parents to take out a loan in order to buy the first of many computers that would fuel is innovative spirit. Decades later, after serving as the Head of Security for the Israeli President and Prime Minister as part of the Israeli Secret Service, Dror went on to found multiple start-up companies. With expertise in fields ranging from 3D and entertainment technologies to programming and robotics, Dror is a serial entrepreneur with a passion for using innovation to create new opportunities. Very much a citizen of the world, Dror has lived in Europe, China and the US, and brings to the table his rich multi-cultural experiences and intimate understanding of Jewish innovation.

**Huang**, a country-raised Chinese native who rose to the challenges and heights of her early education, was awarded many scholarships and earned multiple top university degrees both at home in China and abroad. Huang has a B.A. and M.A. from Peking University and Ed.D. from Columbia University. Throughout her career, Huang has excelled as a cross-cultural arbitrator and has gained a deep understanding of the international corporate and education world and its inner-workings. After living in NYC for ten years, Huang returned to China and founded an education consultancy firm working with high net-worth Chinese families on their children's education planning. Now residing in Shanghai, Huang is renowned as a successful entrepreneur and leader in international education.

Besides both having resided in Shanghai and working in the education industry, what is it that brings our authors together for the purpose of writing this book? Both Dror and Huang are motivated by a central, mutual belief: Learning from the Jewish principles, traditions and education systems that resulted in being awarded 22% of Nobel Prizes is vital in our quest to help Chinese students become successful, internationally recognized innovators. As such, children, parents, and teachers in China must have access to these insights and tools. With their eyes looking towards the future, both Dror and Huang understand that it is by examining the past that we can create a new and improved global society for generations to come.

While the content presented is practical in nature, this book was developed and co-written as a work of the heart. It offers a unique, cross-cultural combination of views regarding Jewish and Chinese innovation. With each topic presented, both proverbial sides of the coin are examined. We seek to balance the history and circumstances of notable Jewish innovators with the current Chinese culture and educational system. Finally, we offer our analysis and interpretation of the past and present in order to provide insights into the "DNA" of innovation. We are passionate about leveraging unique expertise and differing experiences in order to provide a guide for the future success of the Chinese nation. In addition to offering help, this book seeks to inspire hope. As self-made, motivated entrepreneurs, we will also present the individual circumstances and obstacles we have each overcome in order to reach success. In the process, we wish to kindle the entrepreneurial spirit of each and every reader, spurring him or her on towards innovating better solutions for the future.

The following chapters will offer an in-depth study of the ancient systems of learning for both Jewish and Chinese cultures, the evolution of the modern education system in both cultures and biographical examples from the lives of Jewish Nobel Prize winners. After exploring the historical elements of both cultures and their respective education traditions and values, we also put forth a series of keywords and educational concepts that can be integrated into the lives of students in China. The conclusions presented at the end of each chapter will offer insights based on the successes of Jewish scholars, academics, and inventors throughout the ages.

# Chapter 1:
# An Exploration of Common Values

While at first glance Chinese and Jewish peoples and their histories seem quite antithetic, many aspects of these two cultures stem in fact from shared values and parallel historical developments. To better understand the convergence of Chinese and Jewish beliefs and practices, we first turn to the past. The rich histories of both peoples have been critical in shaping the respective customs and present-day daily practices of Chinese and Jewish individuals, families and communities.

In the case of the people of China, these traditions date back to over 1600 BCE and have evolved and developed along the Yellow River and the Yangtze River Valleys. Considered one of the oldest civilizations, China's multiple kingdoms and dynasties developed a culture, literary excellence, and advanced philosophies rivaled by no other for their historic time period. The Chinese people have a long and illustrious history that remains highly respected to this day. Similarly, religious literature tells of the founding of the Israelite nation as far back as 1500 BCE. Jews lived in the fertile crescent, in the land between the Mediterranean Sea and the Jordan River. From their early beginnings they were constantly ruled over and many times marginalized by conquering empires including the Babylonians, Greeks, and Romans. Despite these challenges, the Jewish people persisted, multiplied, and grew stronger in both their faith and their culture of scholarship, leadership, and learning. Much like the people of China, Jews have long prioritized their loyalty to ancient Jewish traditions and beliefs, often fighting wars and losing lives in the name of preserving these traditions and values. While Jewish and Chinese cultures have advanced academically, philosophically and socially throughout the years, both have invested extensive time and resources, generation after generation, to maintain the cultural traditions of their people in order to ensure that their core values remain intact.

One of these core values that has remained central to both Chinese and Jewish cultures is the importance of education. For countless generations, Chinese and Jewish families have prioritized a culture of learning, often going to great lengths to ensure that their children have access to the best academic institutions. In China, from the time of the dynasties and through to today's "China Dream", students are urged to become successful through their academic excellence. There is a strong focus on learning and each child's responsibility to leverage their studies in order to better their lives and the lives of those around them. Similarly, Jewish families both in Israel and around the world have traditionally invested extensively in the education of their children. A 2016 report from the Pew Research Center found that nearly all Jewish adults ages 25 and older around the world (99%) have at least some primary education, and a majority (61%) has post-secondary degrees. When exploring this focus on education among both Jewish and Chinese cultures, it is notable that the concept of "tiger moms," or mothers who serve as the main advocates and opportunity-seekers for their children, is extremely common among Jewish and Chinese families. While some aspects of the stereotypes around "tiger moms" may stray from the truth, both Jewish and Chinese mothers are known to have extremely high expectations for their children, particularly with regard to education. The focus these mothers have on success has led to many high-achieving children. Many of the ideas explored in further detail throughout the remainder of this book rely on the fundamental understanding that both Chinese and Jewish cultures highly value educating future generations and establishing standards of academic excellence among the youth in their communities.

Perhaps no less important is the value placed on family in both cultures. A well-known Chinese proverb perhaps provides the best insight: 家和万事兴 – which in English translates to: "if a family lives in harmony, everything will prosper." Until fairly recently, it was rather standard for up to four generations of a Chinese family to live under the same roof, acting in many ways as one entity rather than separate family units. While the landscape of the Chinese economy and population has undergone immense change over the past several decades, it is still quite common for grandparents to play a significant role in the child rearing of their grandchildren, enabling their own children to remain focused on career advancement and providing for the family. Similar to the way education and childrearing has always been a family-run affair in China, Jewish families have long-standing traditions and practices that act like glue in preserving family life. From family gatherings around Sabbath dinners and holiday celebrations to unwritten cultural norms wherein grandmothers are the assumed caretaker for their grandchildren when both parents are unavailable.

The shared values of Chinese and Jewish peoples often stem from the tomes of writing and philosophy which filled the libraries of ancient times. Both cultures take pride in their written and oral traditions and many overlapping principles stem from these ancient beliefs. One of the most prominent examples of this is what is referred to as the Golden Rule: Do unto other as you would have them do unto you. This belief developed independently and in parallel within both cultures. In Judaism this "ethic of reciprocity" comes from the Talmudic saying by Hillel the Elder, "That which is hateful to you, do not do to your fellow" (Shabbat 31:1), a principle which is said to summarize the essence of Judaism. The Chinese equivalent is attributed to Confucius: 己所不欲，勿施于人 "

勿施于人 which translated into English reads: "What you do not want done to yourself, do not do to others." (James Legge translation, "Analects," 15:23).

The Golden Rule is only one of the many values derived from Jewish religious figures, texts and principles. Perhaps the most notable and well-known set of principles is the Ten Commandments. As the Biblical tale describes, the Jews escaped slavery in Egypt were led by Moses as they wandered the desert between Egypt and Israel in search of their homeland. During these years, Moses received a message from God telling him to ascend to the top of Mount Sinai. Moses obeyed, and there he received two stone tablets from God, each with five laws carved into it – what became known as the Ten Commandments. These Ten Commandments are in many ways the cornerstone of the laws of the Jewish people. They have since been translated into common English as a Decalogue or set of 10 rules that have been adapted by many cultures around the world. Below is the list of the Ten Commandments as well as the excerpt from the Holy Books from which they are taken:

1. "I am the Lord your God, who brought you out of the land of Egypt, from the house of slavery." (Exodus 20:2)
2. "You shall not recognize other gods before Me. You shall not make for yourself a carved image, or any likeness of what is in heaven above or on the earth beneath or in the water under the earth." (Exodus 20:3–4)
3. "You shall not take the name of the Lord your God in vain, for the Lord will not leave him unpunished who takes His name in vain." (Exodus 20:7)
4. "Remember the Sabbath day, to keep it holy. Six days you shall labor and do all your work, but the seventh day is a Sabbath to the Lord your God; you shall not do any work, you or your son or your daughter, your male or your female servant, your animal or your stranger within your gates." (Exodus 20:8–10)

5. "Honor your father and your mother, so that your days may be prolonged in the land which the Lord your God gives you." (Exodus 20:12)
6. ."You shall not murder." (Exodus 20:13)
7. "You shall not commit adultery." (Exodus 20:13)
8. "You shall not steal." (Exodus 20:13)
9. "You shall not bear false witness against your neighbor." (Exodus 20:13)
10. "You shall not covet your neighbor's house; you shall not covet your neighbor's wife or his male servant or his female servant or his ox or his donkey or anything that belongs to your neighbor." (Exodus 20:14)

A central part of religious life, Jews can spend years learning the ways in which to follow and honor the Ten Commandments and the many lessons found in the *Torah*, or holy book. They use the words of the Torah as a guidebook for their daily lives as Jews. It is important to note that Judaism is viewed as a way of life that extends beyond a set of religious beliefs. The written and oral traditions of Judaism are passed down from one generation to the next and both have a special place in each Jewish individual's upbringing and education. There are six central tenants, in addition to the Ten Commandments, that act as guidelines and core beliefs which shape Jewish life and Jewish learning. Jews across the world, no matter where they are raised, are familiar with these principles. It is notable that each of these six tenets has an equivalent in Chinese culture, further exemplifying the shared values between the Chinese and Jewish peoples.

The first tenet prescribes to always do good – transliterated from Hebrew as *ma'asim tovim*. Judaism encourages individuals to find opportunities to bring goodness into their lives and the lives of those around them. Young children grow up learning the importance of simple acts of kindness such as greeting parents and siblings with cheer in the morning. The importance of good deeds is equally revered in Chinese culture. Traditional Chinese teaching values focuses on: 与人为善 - or helping others do well. Likewise, according to Confucian Analects, there is the important concept of: 君子成人之美，不成之恶, which translates in English to, "a gentleman helps fulfill others' cherished hopes; he does not help evil deeds."

The second tenet is acting with kindness – transliterated from Hebrew as *gemilut chasadim*. By being compassionate to others with no expectation of anything in return. From feeding the hungry to visiting the sick, Jewish rabbinic teachers articulate the importance of doing *gemilut chasadim* throughout one's life. In addition to the Golden Rule which refers to compassion and respect for others, the Chinese people believe: 日行一善，积善积德，or in English: perform a good deed every day, accumulate kindness and virtue. According to Lao Tzu's Tao Te Jing, the best virtue of man is like water，which benefits all things, but strives for nothing, which in Chinese is written: 上善若水，水善利万物而不争。

The third tenet is that of hospitality – transliterated from Hebrew as *hachnasat orchim*. This is considered a major part of Jewish life. Inviting family, friends, and even strangers into your home can teach individuals the importance of treating others graciously and acting generously towards others. In Chinese, hospitality is written with two characters: 好客. The first character, 好, symbolizes affection or love, while the second character, 客, translates to guest. In that sense, the very nature of the word "hospitality" translates to showing affection or love for one's guests. Much like in Jewish culture, hospitality plays a very important role in Chinese daily life. Perhaps the role of hospitality in China can best be described by a famous proverb: 有朋自远方来，不亦说乎，which translates in English to "it is a delight to have friends from afar."

The fourth tenet is charity, transliterated from Hebrew as *tzedakah*. The concept of *tzedakah* is very prevalent in daily Jewish life and is there to reminds individuals that they should be generous and give what they can to charitable organizations and individuals in need. The Chinese phrase: 勿以善小而不为 mirrors this concept, translating to "do not fail to do good even if it is small." The Chinese believe that helping others will benefit the individual who does the helping as well. One of the memorable Chinese phrases demonstrating this principle is: 赠人玫瑰, 手留余香, translating in English to: "roses given to others leave a fragrance in your hand."

The fifth tenet is visiting the sick, transliterated from Hebrew as *bikkur cholim*. This principle is designed to instill in Jews the need to help others heal. From making sure that elders are comfortable to assisting those who are ill and cannot help themselves, the concept of *bikkur cholim* reflects the way Judaism seeks to promote respect between individuals. As with the other tenets, there is a similar principle within Chinese culture expressing this value. The Chinese phrase: "老吾老以及人之老, 幼吾幼以及人之幼" translates in English to: "extend the respect of the aged in one's family to that of other families. Extend the love of the young ones in one's family to that of other families."

The sixth and final central tenet is called bad language, transliterated from Hebrew as *lashon hara*. This tenet is one of the ways in which Judaism teaches individuals to think before they speak and to strive to never speak ill of others. The Torah teaches Jews not to embarrass others in public, not to lie, and to understand the weight of words - to know that they can be just as hurtful as they can be helpful. This principle is equally important in Chinese culture and values, demonstrated clearly with the phrase: "良言一句三冬暖, 恶语伤人六月寒," translating in English to mean: "a good word brings warmth to people through three winters, evil language on the contrary hurts people and makes June seem as cold as the winter."

Despite the many convergent values, there are also several examples of the way shared principles were applied differently within Jewish and Chinese cultures. One such example is the way dishonor and obedience play a specific role in Chinese culture and play a very different role in Jewish culture. As we have seen through both the religious principles and cultural norms central to Chinese and Jewish daily life, respect is a core value that shapes the beliefs and actions of each individual. As such, it is no surprise that both Jewish and Chinese cultures have a strong concept of what it means to act dishonorably or to disrespect one's culture or faith. Whether one's actions dishonor their God, family, party or nation – Chinese and Jewish people know well the cultural repercussions of dishonorable behavior and go to great lengths to avoid such scenarios. In China, wanting to avoid such dishonor is often used as a way to encourage and enforce desired behaviors with regards to the education and study habits of Chinese children. Within Jewish communities, this same strategy was less commonly found within the education system, where questioning and rule-breaking is accepted, and instead mostly centered around encouraging individuals to maintain their relationship with God and remain committed to living a life centered around Jewish values. As we explore these concepts in further detail in later chapters, we will address the divergence in the application of shared values, the core values that remain unique to each of the two cultures, and the way each serves as a puzzle piece, shaping Chinese and Jewish cultures into what they are today.

# Chapter 2 : What is Innovation?

**Imagination is more important than knowledge. For knowledge is limited, whereas imagination embraces the entire world, stimulating progress, giving birth to evolution. It is, strictly speaking, a real factor in scientific research.**

*-Albert Einstein*

Pete Foley, behavioral scientist, defines innovation as "a great idea, executed brilliantly, and communicated in a way that is both intuitive and fully celebrates the magic of the initial concept." We need all of these parts to succeed. Innovative ideas can be big or small, but breakthrough or disruptive innovation is something that either creates a new category, or changes an existing one dramatically, and obsoletes the existing market leader." In the field of economics, management, and practice and analysis, innovation is described as "the result of a process that brings together various novel ideas in a way that they affect society." And further, for business and consumer purposes, innovation is defined as "the application of better solutions that meet new requirements, unarticulated needs, or existing market needs. This is accomplished through more-effective products, processes, services, technologies, or business models that are readily available to markets, governments and society."

The question we pose here is **how**. How does an individual or a company become innovative? If every Jewish Nobel winner can be described as adding an innovation to our world, how did they become innovative? What factors motivated them throughout their early childhood, studies, and into adulthood? What does innovation require? How can it be achieved? These questions and more will be evaluated in the following chapters. But first, we present a 'diffusion of innovation' graph to define specifically what aspect and area of creativity and success we are measuring.

**Diffusion of Innovation**

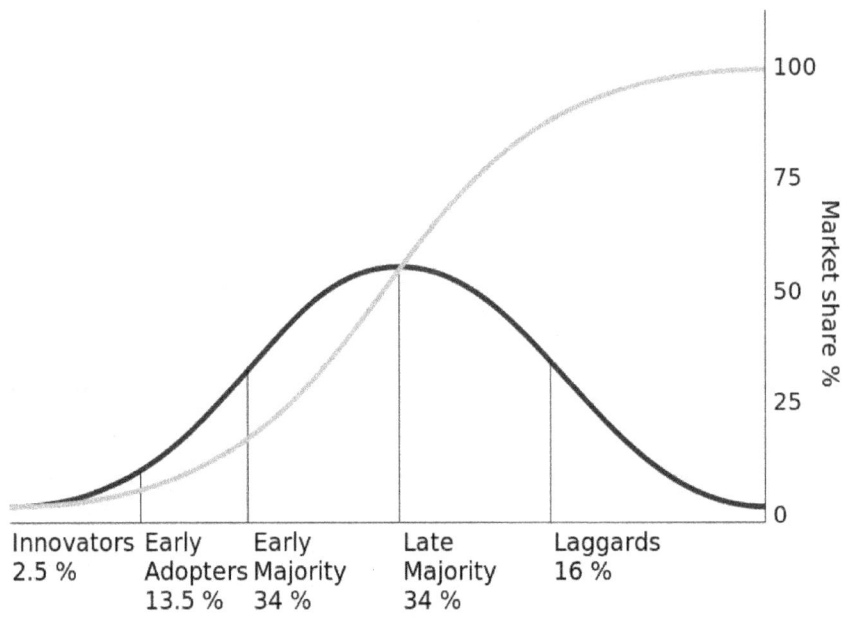

Popularized by communication professor Everett Rogers, this Diffusion of Innovation graph presents the five main stages of innovation and how it is diffused within the market and eventually widely adopted across the market. As is widely acknowledged and as we will show in the chapters to come, Jewish people have often been the harbingers of innovation. From Nobel Prize winners to modern Israeli start-ups, Jewish scientists and engineers have shown their propensity for disrupting markets, coming out as technology leaders and serving as the innovators and early adopters in the graph above.

The need for strong innovators leading our society and addressing global challenges has never been more critical. As described by President Clinton's former Secretary of Education, Richard Riley, "the jobs in the greatest demand in the future don't yet exist and will require workers to use technologies that have not yet been invented to solve problems that we don't yet even know are problems." In order to prepare for this future, educating our students today should focus on nurturing innovation. Students should be able to combine curiosity and critical thinking in order to problem-solve their way through obstacles. As a society, we must strive to keep creativity at the center of the learning curriculum. Unfortunately, most students in today's education system are severely lacking in these critical skills. While we can all agree that there are certain topics and subjects that will always need to be taught in schools, we must find ways to cultivate an innovative mindset among students. Only be fostering a sense of innate curiosity and questioning can we prepare current and future generations for, as Riley put it, "solving the problems that we don't yet even know are problems."

Respected *New York Times* columnist Thomas L. Friedman, author of *"The World is Flat,"* has set forth a formula that captures the importance of curiosity and innovation:

$$CQ + PQ > IQ$$

Curiosity Quotient     Passion Quotient     Intelligence Quotient

Friedman explains, "Give me the kid with a passion to learn and a curiosity to discover and I will take him or her over the less passionate kid with a huge IQ every day of the week." IQ of course still matters, according to Friedman. But curiosity and passion matter even more. It is curiosity and passion that spark innovation.

It is this innovative thinking process that can open up a new world of opportunity for today's Chinese students and generations to come. While there are clearly individuals in China who have already subscribed to Friedman's theorem above and in addition to their impressive IQs are also curious thinkers and passionate learners. However, these individuals may be considered outliers in today's China. We propose that such outliers can and should represent the new standard of success for Chinese students. With the help of parents and educators alike, Chinese students can move from traditional, exam-focused learning to a path of discovery that fuel's students' passion and curiosity. In doing so, China will create what we will refer to as the "cultural infrastructure of innovation."

**History of Innovation: Old and New**

One could argue that the Jewish people have been practically forced into becoming innovators due to their history of political, social, and cultural marginalization. When given no other choice, one must create new circumstances for oneself – essentially, innovating to create new options. While it may not be the only historical catalyst for innovation among Jews, it is at the very least a significant motivator. However, as history proves, not every group of people in every culture that faces a lack of choice was able to innovate and create new opportunities for themselves. There are a number of additional factors and cultural experiences that aligned to create the previously mentioned "cultural infrastructure of innovation."

While innovation among the Jewish people dates back centuries, perhaps this cultural infrastructure of innovation is best demonstrated by looking at some examples of Jewish innovation in Europe and how Jews positioned themselves as became banking experts and lending innovators. During the Middle Ages, Jews throughout Europe were confined to certain areas, forced to live outside of the city walls and were not allowed to own land. Throughout most of Europe they were persecuted and had to abide by strict rules and regulations. Forced to make a living outside city walls across Europe, the Jewish people developed a tradition of trading and bartering from the grand silks and spices from the Orient. So how did these Jews, without any land or power, become Europe's top banking experts and enable their people to not only survive, but attain a certain degree of freedom? Simply put – they recognized an opportunity, became innovators and created an industry of their own.

Around the 14th-16th century, Christians in Europe were forbidden to practice usury, or charging interest on loans. Jews recognized that while Christians could not lend with interest, they certainly could! This marked the launch of Jewish lending businesses and systems that went on to control much of the international grain trade of the time. In addition to betting and hedging on future crops, Jews also began leveraging their place in the international market in order to finance some of the greatest families of the Renaissance in their art, cities, and armies. Jews across Europe soon became invaluable to the ruling classes and played a critical role for powerful kingdoms for years to come. In examining this history, we can see that the innovative spirit of the Jews in Europe arose not in spite of being mistreated or despite not being allowed to own land, but <u>because of it.</u> They also had to be savvy enough to recognize a unique opportunity that no one else had yet pursued, or, in this case, no one else was allowed to pursue. It was when faced no other way out, a need to secure their livelihood and an untapped opportunity that Jews in Europe sought out an innovative approach to improve their lives and founded the European banking and lending industry in order to create a new reality for themselves, their families and their communities.

Chinese history is certainly no stranger to innovation. Credited with range of critical inventions such as paper making, gunpowder, the compass and movable engraved printing, Chinese innovators from across various lands and dynasties created solutions to important problems they faced in their daily lives. As is the case with many inventions, Chinese innovators were able to create a new solution to a well-known challenge.

Let us take for example the paper-making discovery made by Cai Lun in 105 AD. This servant of the court defined the problem - all written materials had to be inscribed either onto heavy bamboo or expensive silks. Cai Lun was able to completely change a far-reaching cultural norm through his invention of paper made out of lighter materials. His innovation made possible the recording, transportation and distribution of mankind's oral histories. In the book "The 100 – a Ranking of the Most Influential Persons in History" by Michael H. Hart, Cai Lun is ranked as the seventh most influential person in history. This list includes Gutenberg, Einstein, Pasteur, Galileo and Aristotle. His place in this order of "The 100" is truly significant. It reinforces the historical context and future potential of Chinese innovators within a China that values scholarship and creativity. It is now that we must ask: How can China create its next Cai Lun? How can China create its next great innovation that will have as far reaching an effect across the globe for centuries to come as the discovery and standardizing of paper?

European Jewish bankers may have been the first Jews to innovate by 'cornering a market' in order to make a living, but they certainly were not the last. This precedent of innovation arising from out lack of other opportunities can also be found in the establishment of what we now call "Hollywood". Hollywood was essentially the realization of a dream. Immigrant Jews in the United States and first-generation, American-born Jews landed on the east coast of America starting at the end of the 19th century and hoped to find a place where the underdog could make a living if he worked hard. Instead, that dream was quickly shattered and replaced with a reality wherein Jews couldn't work in finance, attend the best schools, own any honorable business or participate in many professions due to strict quotas and anti-Semitic policies. Many Jewish families found themselves relegated to immigrant ghettos, not so very much different from their living conditions in Europe. It wasn't long before a number of them decided to try to 'make it' elsewhere in the United States. They left the 'new' Jewish ghettos of the Eastern port cities and headed to the West coast.

The first moguls of Hollywood were professional salesmen. They approached the motion picture industry in the same way they approached the sale of other items they had sold successfully in other industries. Moving from clothes, gloves, diamonds and furs to the movies – these salesmen saw in the budding movie industry the possibility to create mass appeal with one important added bonus: the customers paid money before they saw the product! Prior to Jews creating the movie industry in the Hollywood hills of California, film was widely considered entertainment for the public working classes. Films would idealize an America where the upper echelon ruled and where immigrants were portrayed only as secondary characters, waiting on the sidelines. Movies were often racist, anti-Semitic, and continued a long history of stereotyping the 'other.' It was essentially a medium for the continued idolization and status quo of a privileged class of white Protestants.

The Jewish founders of Hollywood were able to become movie innovators by simply integrating their traditions of lifelong learning, appreciation for culture and different forms of artistic expression into the movie industry. Having been shut out of a range of upper class professions in America, Jewish movie producers came into the industry with a deep motivation to succeed and the grit needed to thrive in a cutthroat business. They understood that if they weren't first to innovate, their movies would lose at the box office. These born-and-bred salesmen leveraged everything they knew from their experiences "wheeling and dealing" as traveling salesmen and as marketers in their efforts to disrupt the movie industry and in the process, created a new class of films aimed at middle class Americans. They applied the same skills that helped them survive in the ghetto in order to create and mass market the American dream to a previously untapped audience.

While there are many similar stories of innovation in China, perhaps the most prominent, modern day Chinese equivalent is that of Alibaba founder, Jack Ma. This familiar, 'rags to riches' story demonstrates how innovation led to establishment of one of the most successful online commerce companies world-wide. Similar to the humble beginnings and lack of resources of Jewish immigrants in the US, Jack Ma, born in Hangzhou, had no money and no familial or business connections to leverage in order to get ahead. The only way that he could create new opportunities for himself was by excelling as a student. Ma finished high school, but despite his best efforts, he failed his entrance exams twice before he was finally accepted to university. By refusing to give up and instead remaining committed to his studies, Ma was able to attend Hangzhou Teacher's Institute. After graduating university, he was rejected by numerous job prospects and took a job at a local KFC chain before he was finally hired as an English teacher. He still had a thirst for learning and pursued an advanced degree. He applied to Harvard University and was rejected 10 times.

How did this young man with little experience and a history of rejection become one of the world's most impressive innovators and early adapters of the computer and coding industry? On a trip to the US in 1995 he was first introduced to the internet. Ma realized that China was ripe for innovation and new opportunities in the internet space. Without waiting until he had a perfect product in hand, Ma immediately gathered the seed money for his venture *China Yellow Pages* and related ventures that would bring corporate and personal website coding to China. These early efforts would turn out to be first iterations of what would become the global powerhouse known as Alibaba. Filling a newly discovered need, Ma was able to beat his competitors to this new e-commerce industry in China and tap into the largest customer base in the world.

Ma himself admits that he did not own a personal computer until the age of 33. We know that his academic career is less than illustrious. Thus, we seek to understand: What enabled a man without advanced degrees or computer engineering skills to create one of the most innovative e-commerce marketing platforms in China? Perhaps more importantly, we must examine and seek to better understand Ma's history and the precursors that led to his company having the largest IPO in the history of the New York Stock Exchange with over US$ 25B raised.

The reality is that Ma's story need not be one of a kind. Other innovators in China can achieve the same level of success and China can work to create the infrastructure of innovation that nurtures determined minds like Ma's. The main challenge China faces in creating this infrastructure is bringing into the classroom a thirst for critical thinking, a propensity for challenging the status quo through questions and creativity and an ability to quickly recognize new opportunities. Innovation is on the rise in China and the *Financial Times* recently confirmed that the speed at which Chinese companies move and adapt is shocking to western counterparts. "[In China], it's more important to have a finished product than a perfect product," writes the *Times*. China is fast becoming a global leader in innovation, leading many recent Western media reports to agree that "it's time to copy China." The world is eagerly watching China's booming businesses and the innovation that made their success possible. Now, we must instill in Chinese students the realization that, like Jack Ma, being first to the market is no less important than meeting a market's needs. Being first to answer the question or solve the problem is often the crux of innovation and it need not be based on creating the "perfect" solution.

One of the most fascinating stories about being first as an innovator centers around the rise of Facebook and its founder, Mark Zuckerberg. Zuckerberg grew up in a Jewish household just outside of New York City. He excelled in school as a young child and eventually transferred to Phillips Exeter Academy, a private preparatory high school in New Hampshire. Throughout his early education, Zuckerberg dominated competitions and won honors and awards in multiple subjects including math, astronomy, classics and physics. He was interested in computers and began programming from an early age. Zuckerberg is said to have worked from his father's home computer where he taught himself basic coding. His parents quickly saw his potential and secured private tutors that could teach him coding and computer technology during high school. As a teenager he was already building software and creating computer games for his friends. Given his abilities, it was no surprise that at 18 he was accepted to Harvard University.

Zuckerberg's time at Harvard was relatively brief. In his second year at the university, he had developed the first iteration of Facebook, at the time called TheFacebook. Less than a year after its initial launch and following a summer of growing the platform making connections in Silicon Valley, Zuckerberg decided not to return to Harvard. Barely twenty years old at the time, making that type of decision took courage and belief in his skill and talents. After many late nights of coding in his college dormitory, Zuckerberg knew he had to be first and stay one step ahead of everyone else. It was with this understanding that he was able to take the leap, drop out of Harvard and pursue the development and expansion of Facebook full-time.

While the original idea continued to evolve with each update and expand to new audiences around the world, it was the belief in his initial idea and the deep understanding that he must be first that made his success possible. Zuckerberg's story is not just one of innovation but of courage to follow his passion and creativity, even when it meant leaving a premier

Ivy League university such as Harvard. According to Time Magazine, Zuckerberg is among the 100 wealthiest and most influential people in the word. In December 2016, he made the Forbes list of the top 10 Most Powerful People in the World. With a net worth estimated at US $ 71.8 billion as of 2017, Zuckerberg is also ranked by Forbes as the world's fifth richest person. But Zuckerberg's success is certainly not only measured by his worth in dollars. He has vowed to 'advance human potential and promote equality' through "The Giving Pledge", a contract that he and his wife, philanthropist and Doctor Priscilla Chan, have signed - marking their intention to give, over the course of their lives, a majority of their wealth to various international causes and humanitarian efforts.

Following the above examples of both Jewish and Chinese innovation that were both 'quick' and 'first' to solve and address a need successfully, below are seven practical and useful ways in which we can learn from these historical examples, incorporate what we know about our current global society and translate this knowledge into lessons for the modern Chinese classroom. These recommended activities, exercises and skills can be used to cultivate the next innovators of China and are designed to be completed both in school and at home.

1. <u>Infuse the Spirit of Innovation into Day-to-Day Tasks</u>: Innovation need not be a new, separate curriculum to be integrated into a child's already busy schedule. In fact, the practice of innovation is perhaps best developed through the integration of creativity into a student's daily life. Start by asking your student to approach one of their everyday tasks in a new way. Ask students how they can find a new and perhaps innovative way of completing their schoolwork, getting dressed, or even cleaning their room. Most importantly - keep asking! Innovation must be practiced daily. When creativity and innovation become the routine, a new mindset can be achieved.

2. <u>Demonstrate the Difference Between Innovation and Problem Solving</u>: While they are often used interchangeably, these terms are not one and the same. Problem solving is often about using existing knowledge to provide an answer to a well-defined question. Innovation on the other hand is based on using creativity to design a new approach, reframe the question and overcome the challenge. Help your student understand the difference by providing examples from their daily lives that demonstrate where they or others have used problem solving or innovation to arrive at a solution.

3. <u>Model and Practice Innovative Behavior as Parents</u>: Parents who would like their students to innovate must become innovators themselves. From a simple change in the way a meal is served or the order in which shoes are arranged, find a reason to innovate. Share it with your child. Give yourselves the opportunity to invent something new and work together to discover innovative solutions. Think of new ways to complete daily tasks more efficiently or productive. Approach these innovation challenges without a fear of failure for we must remember and teach children that innovation requires a willingness to fail.

4. <u>Dive Deep into Learning and Ask Questions</u>: Chinese students are trained to be ready to answer any question when asked, but in order to cultivate innovation we must allocate time for students to ask themselves questions and deepen their learning experience. Not sure where to start? Perhaps there is an aspect of something your student learned at school that they want to know more about, or don't quite understand. By enabling him or her to ask questions and express uninhibited curiosity, you can develop his or her ability to think deeply and critically about the knowledge they have acquired. Use positive feedback methodologies such as encouragement charts and physical rewards to link the experience of asking questions and deep learning with feelings of enjoyment and pleasure.

5. <u>Eliminate the Fear of Failure</u>: Fear of failure is the #1 killer of curiosity and innovation. It is essential that students know that they can explore and learn without the fear of asking a "wrong" question or failing to create a viable solution. The sooner students accept that failing is a critical part of the innovation process, the sooner they can become accustomed to this reality rather than fear it. By encouraging rather than disciplining students for asking questions and offering up ideas, regardless of if they are correct or "perfect", schools and parents can eliminate a central barrier on the path to becoming an innovative thinker.

6. <u>Develop Computational Thinking Abilities</u>: Computational thinking is a way of solving problems that draws on concepts fundamental to computer science. To flourish in today's world, computational thinking has to be a fundamental part of the way students think and understand the world. It is perhaps one of the most critical 21$^{st}$ century skills as it will enable students to address the increasingly complex challenges we face as a global society. From music and art to robotics and coding, expose your student as often as possible to the tenets of computational thinking such that they can be integrated into his or her way of approaching new problems.

7. <u>Ignite Passion</u>: The last but perhaps most critical lesson we can learn from historical examples of innovation is the importance of passion. Students must be genuinely excited and intrigued as they learn. Rather than feeling like a "cog in the wheel" within a vast educational machine, or simply focusing on acquiring new facts, students must be inspired! They must literally light up the "wires" in their brains with new connections and ideas. This excitement for learning must be found in everyday subjects. Next time your student shows genuine intrigue about something they are learning, encourage them to explore further. Create opportunities for interactions with individuals that are passionate about what they do. When individuals are driven

by passion, one's entire body – including both the mind and heart – are engaged in the innovation process.

In the following chapters we will continue to delve deeper into the precise elements found within the "infrastructure of innovation." We will explore and uncover the components of the creative spirit found within individuals whose names are thought to be synonymous with the word innovation: Nobel Prize winners, the faces of today's modern tech industry as well as those that were able to use their creativity to change the course of history.

# Chapter 3:
# From Survival to Success –
# The History of Modern European and American Jewry

The ancient history of the Jewish people from the Old Testament is familiar to most students and followers of similar monotheistic religions. However, the more modern and contemporary history of the Jewish people may not be as well known, particularly in places throughout the world without a significant Jewish population. Throughout history, Jews have been forced to leave their homelands and persecuted for their religious beliefs. The term "diaspora" is used to describe Jews living outside of the land of Israel, otherwise known the Jewish homeland. The Jewish Diaspora includes the many communities of Jewish people scattered around the world. As mentioned in the previous chapter, in the Middle Ages, Jews were banned from living in major cities and instead settled mostly on the outskirts of towns. They were not allowed to own or conduct businesses in the majority of the Christian nations where they were living. Despite the many obstacles, Jews used education and innovation to leverage whatever advantages they could and managed to both maintain and grow their communities. This was particularly true among the European Jewish diaspora. However, what began as marginalization and discrimination by Christian leaders soon became the most existential threat to the Jewish people in modern history – the mass extermination of Jews in Europe that later became known as the Holocaust.

The rise of Adolf Hitler and the Nazi party in Germany in the early 1930s was the result of a multitude of social and economic issues facing the country and Europe as a whole. The common thread of the beliefs and goals of the Nazi party and the propaganda they published throughout the years was the restore the wealth and power of the Aryan (aka: White, Christian) German and eliminate were referred to as the "second-class races" and religions around them. Using propaganda that incited racism, antisemitism, and classism, the Nazi party grew stronger. Operating as a dictatorship led by Hitler, the Nazis fueled their rise to power with the claim that the economic troubles of Aryan Germans were a result of the "evils of the foreign Jews that had infiltrated Europe." To support these claims, they pointed to the Jews throughout Europe that, based on their innovation and academic excellence, had become leaders in finance, commerce, the press, literature, theater, and the arts. Nazis attributed the downfall of the modern German Aryan citizen and the economic hardships that they faced following World War I to the success of the Jewish people in these various sectors of German and European society. Using this racist and anti-Semitic propaganda, the Nazi party began building an entire system through which to remove and exterminate all Jews living within the Aryan empire they sought to establish across Europe.

Arguing that only individuals of "pure" German descent should be allowed to serve as leaders in German society, the Nazis began removing Jews from businesses, universities, and the arts. They made every Jewish person wear a yellow star on their clothing as a way of identifying them. Jews were soon forced from their jobs, their homes and their communities and sent to live in what became known as ghettos. Jews were expelled from the same society and culture to which they had contributed as lawyers, doctors, bankers, scientists, engineers, professors, writers, publishers, artists, and musicians. They were denied access to their own bank accounts and all their possessions were turned over to the Nazi regime. Many Jewish families tried escaping to neighboring European countries that had not yet been taken over by the Nazi party. Others tried fleeing the United States, Canada, and South America. However, even immigrant-based countries such as the USA instituted harsh quotas on Jewish immigration, turning away hundreds of thousands, if not millions of Jews seeking safety. It was not long before the Nazis began implementing the atrocities of what they termed the "Final Solution" – the mass murder of Jews throughout all Nazi-held territories.

Concentration camps and labor camps where these planned mass executions would take place were built on vast land holdings of Eastern Europe which were now under Nazi control. Jews were gathered city by city, town by town, and forced to march to their death. If they made it to the concentration camps, Jewish men, women, and children were methodically killed by poison gas within a few hours of their arrival. They were unloaded from overcrowded trains, separated by age and gender, forced to undress, their possessions were taken away and looted, they were killed with gas showers, and the bodies were taken immediately to special mass crematoriums. Others were forced to dig their own mass graves on the side of the road and stand inside waiting to be shot in the thousands. Throughout this terrible and unbelievable atrocity that would later be termed the Holocaust, many nations stood by, turning a blind eye or claiming ignorance. According to research by Sergio Della Pergola of the Hebrew University of Jerusalem, in 1939, 9.5 million Jews lived in Europe. At the end of World War II in 1945, the Jewish population of Europe was only 3.8 million. Most estimates by historians and research studies agree that somewhere around 6 million European Jews were killed during the Holocaust.

In reviewing and analyzing these terrible historical events, we must ask: How did the Jewish people go from losing over one-third of their global population to becoming so successful throughout the world? What effects did the Holocaust have on the Jewish people that led to so many of its survivors finding ways to invent, create and innovate? These questions play an important role for those studying the history of the Jews and their successes throughout the ages. While Jews today no longer face this same threat, the spirit of survival that emerged from these dark times lives on in the decades and generations that followed. The resilience, drive and fight to succeed emanates not only from each individual's goals but from the collective memory of all those who were killed in the Holocaust. This is a central part of the Jewish experience in the second half of the 20th Century.

Between the rise of the Nazi Party in the early 1930s until the end of World War II in 1945 many notable Jewish persons with European backgrounds would flee and become refugees. Perhaps the most notable among them was none other Albert Einstein. Arguably the greatest mathematician and physicist of the time, Einstein could have just as easily perished in the Holocaust. Trying to calculate the contributions that could have been made by Jews that were killed during the Holocaust is unimaginable. Thankfully, of the Jews who did find refuge in other countries or were able to somehow escape deportation and death, many would go on to become educated, successful and renowned innovators. Needless to say, the Holocaust had a sobering effect on the European Jews who survived and those living in other places around the world. However, losing such a large percentage of the global Jewish population seemed to strengthen the resolve of those that remained to do whatever it took to keep their small nation alive and growing.

We need not look far for examples of Jews that, following World War II, went on to succeed as leaders and innovators across diverse sectors. From an education perspective, the 2016 Pew Research Center study on

Religion and Education Around the World found that, globally, Jewish men and women each have 13.4 years of schooling, on average, and 61% of both men and women have post-secondary degrees. 75% of Jews in North America have higher education in comparison with 40% of non-Jewish populations. Financial success among Jews in the diaspora was no less impressive. In a 2016 Pew Research Center report on income variability among populations in the US, 44% of American Jews surveyed reported having a household income exceeding $100,000, compared with only 19% of adults surveyed in the US public as a whole. Studies conducted between 2000 - 2009 report that Jews comprised more than 25% of the people on the Forbes Magazine's "List of the Richest 400 Americans". This number has since risen to 32%. Similarly, 45% of the top-40 "Richest Americans" list is of Jewish descent, and 1/3 of all American multimillionaires are Jewish.

In 2011, Richard Lynn published one of the most comprehensive studies of the Jewish population in the United States, "The Chosen People: A Study of Jewish Intelligence and Achievement." The book uses an 'achievement quotient' based on overrepresentation to quantify Jewish achievement across different sectors. With Jews making up approximately 2% of the total US population, the following numbers help demonstrate the unprecedented success of the Jewish population of the United States:

**Jewish Achievement Quotients for Different Professions, 1960**

| Occupation | Achievement Quotient |
|---|---|
| Psychiatrists | 5.8 |
| Mathematicians | 3.8 |
| Doctors | 3.7 |
| Writers | 3.4 |
| Lawyers | 3.3 |

Likewise, Lynn's research points to a survey conducted by Harriet Zuckerman in 1969 with results that indicate that Jews were approximately three times overrepresented among university faculty and seven times overrepresented among university faculty in elite colleges. As noted in our introduction, Jews are highly overrepresented among Nobel Prize Laureates. As of 2011, 200 Americans had been awarded the Nobel Prize and of those 62 (31%) were Jewish. This trend of overrepresentation extends into the performing arts and is perhaps best reflected in the high percentage of Jewish musicians, conductors and composers within leading music ensembles and orchestras throughout the US.

These numbers further strengthen the impressive history of success and innovation found among Jewish communities in the biblical homeland of Israel and across the diaspora. This is particularly true when acknowledging the historical context from which Jews were able to strive and innovate. In order to better understand how this success and propensity for innovation was achieved, we will turn our focus to the specific educational methodologies, cultural traditions and norms that make up the infrastructure of innovation found in Jewish communities across the globe.

# Chapter 4:
# Passion, Creativity & Innovation in Early Education

Contrary to what one might expect from a nation with such a strong military and where army service is mandatory, Israeli and Jewish culture is fueled by emotion in general and passion in particular. Modern Israeli society was built by passionate and dedicated pioneers, many of whom risked their lives working tirelessly to establish the infrastructure that would enable this young nation to grow. From finding a way to farm on the arid desert land to teaching new immigrants modern Hebrew such that there would be a national language, passion was perhaps the only resource of which Israelis seemed to have an endless supply. Judaism, both religiously and culturally, has numerous traditions that encourage the expression of one's emotions. Unlike cultures where "wearing your heart on your sleeve" has a negative connotation, in Israel and in Jewish communities across the world, passion is seen as a way to propel new ideas forward and achieve greater success.

Passion is perhaps one of the key ingredients to Israel's success as a hub of innovation and why it was dubbed "Startup Nation" and is home to the highest number of startups per capital. Rather than filtering out emotion in the workplace, Israeli professionals are encouraged to pursue their ideas whole-heartedly. This intrinsically-passionate attitude often leads to a greater sense of commitment and a willingness to explore every potential path to success. Entrepreneurs across Israel tend to operate from a place of instinct and emotional drive, creating companies where intensity and passion are prioritized over skill-level and experience.

Where other cultures may focus on practice and hard work as the key to success, the Jewish people focus first and foremost on cultivating their passion and creativity. In addition to the success of Israeli startups, the central role passion plays in Jewish culture can be seen through the interdisciplinary approach of many Jewish Nobel Laureates. Many of these innovators leveraged their knowledge and process of discovery across a variety of subjects to maximize the impact of their work. Fueled by a deep curiosity, many of these Nobel Prize winners found inspiration somewhere unexpected and found that it paid off to explore new ideas about which they were passionate.

Perhaps the most important take-away for Chinese educators, parents and students is that passion can play an important and positive role on the path to success. From picking up a new musical instrument to learning a new language, doing so with passion can completely change the learning experience and the ability to master this new skill. Passion must come from within but is often ignited by those in the student's environment. This is where teachers, mentors and parents play an important role. Integrating fun, engaging activities into a strict schedule of practice or study can spark students' passion during the learning process. Maintaining the natural curiosity that is particularly strong among young students is perhaps even more important than reinforcing traditional learning techniques. Discipline and strict boundaries can often dim passion and early interest in a new subject. Likewise, endless practice in order to master a skill without infusing this practice with passion can often lead students to feel burned out, stuck or without motivation to continue advancement. Take, for instance, learning to play the violin. If a student has no passion while playing, no matter how skillful it is, it will be nearly impossible for them to truly move an audience or to reach their full potential. A student might achieve high scores, win competitions and awards, but without passion they will never reach new levels of creativity and innovation.

In order to help harness and cultivate students' potential using passion, creativity and innovation, many educators and parents in Israel and around the world have turned to new, alternative educational methodologies. While the vast majority of students in Israel are students in traditional classrooms, over the past two decades Israel has begun adapting a variety of holistic learning methods. With their eye on the future and the need for developing the innovators of tomorrow, the most popular of these approaches are those focusing on cultivating creativity and independent thinking among young students.

The first of these methodologies is the Montessori method, developed in Rome in the 1900s by Maria Montessori. The central tenet of this educational approach is the cultivation of self-learning, with teachers serving as a guide in students' process of exploration. Montessori schools allow students to learn at their own pace through special, self-corrective tools. For example, if a student is trying to put together a puzzle, they will know whether they have done so correctly based on the pieces fitting together to create a shape or figure they recognize, rather than an adult or teacher guiding them and showing them the correct answer. The Montessori program stresses that teachers should assist the students only insofar as engaging them in different activities. Classrooms are often mixed ages, allowing older students to model behavior for younger students and encouraging students to learn from one another. Teachers are present to provide positive reinforcement and ignite students' passion and curiosity, but ultimately children must acquire skills themselves through individual learning. In Israel today, there are just over thirty Montessori method schools of mixed ages beginning in preschool and continuing through grade school. Educators from the Israeli Montessori

Association believe that this style of education, when adapted to the Israeli teaching curriculum can produce the next generation of innovators for companies like Google, Facebook, and Amazon.

An additional alternative educational methodology that has gained popularity in Israel is called the Waldorf method. Also known as Steiner education, this approach is based on a pedagogy of imaginative learning. This method, based on theories of child development, begin with hands-on activities and play as young children, move to artistic expression and social capacity in early education and then focus on critical reasoning and understanding within secondary education. The Waldorf method teaches students about their own individualism and proposes that students should be taught how to think, not what to think. What is especially important in a Waldorf education is the holistic growth of the student as an individual, rather than just on an academic level. Imagination, art and time outdoors are all part of the learning process and that students should feel supported, encouraged and loved, much the way they would in their home environment. As of 2015, there are over 120 Waldorf schools (preschools, elementary schools and high schools combined). Each of these schools advocates for a holistic vision of the student and encourages the development of creativity and individual thinking in each child.

The third alternative learning methodology that has a slightly smaller but still significant presence in Israel is the Reggio Emilia approach to learning. This method was formulated by psychologist Loris Malaguzzi following World War II in the Reggio region of Italy. Malaguzzi's goal was to show that through respect, responsibility, and community, students can become better citizens of the world and discover their strengths through a self-guided curriculum. Students in Reggio Emilia schools are encouraged to explore, and the approach is best known for its project-based approach to learning. Lessons are solely based on the interests and passions of the students. Educators leverage the student's topic of interest to guide him or her through project-based learning modules. If a classroom is interested in insects, a unit on insects will be explored. If another classroom is interested in cooking - an in-school restaurant will be created. Everything is based on the creativity and collaborative discovery. When students are able explore what they are passionate about, innovation and creativity are able to blossom. Reggio Emilia teachers are considered co-learners and are encouraged to learn side by side with the students rather than stand at the front of the classroom. Teachers are expected to ask questions, encourage each student's individuality and serve as a partner in the innovation process.

An interesting aspect of Reggio Emilia education is that schools and teachers are encouraged to avidly document the learning process. Parents and future educators can then leverage teacher logs and observations, photos and videos in order to measure achievements and progress. In Israel there are a number of schools that use the Reggio Emilia method to foster creativity in their students. This is one of the teaching styles that is also sometimes incorporated into public school education for individual classes in early education.

Chinese parents and educators can note that while each of these educational methodologies stress different areas of development for the child, they share the central goals of incorporating creativity into the learning process, instilling a sense of passion for and curiosity, and letting the child have agency in navigating their education experience. These methods can often feel at odds with traditional approaches to education in China. However, Chinese educators and parents need not completely abandon their traditional style of education in order to both learn from and integrate aspects of these alternative methodologies into the daily experiences of Chinese students. Below are some of the core ideas that can be adopted from each of the approaches described above.

Self-learning – The Montessori method has proven that children are highly capable of developing self-learning as a skill that can be applied to all areas of their lives. This is particularly important when considering the realities of our technology-driven world, wherein individuals will be required to continuously adapt and interact with new technologies. Those who are capable of self-learning new skills, both as children and as adults, will gain a unique advantage on their path to success.

Developing a student's "inner compass" – Unlike traditional Chinese education methods where students can feel compelled to think in a particular way or abide by strict rules, the Waldorf approach enables individuals to find their own way. As such, students' individual journeys of discovery leave space for creativity. Finding opportunities for students to use their imagination and encouraging them to explore can help nurture this inner compass.

Letting Students Lead the Way – The Reggio Emilia approach allows educators to reframe the learning process such that students can explore based on their interests and passions rather than a pre-determined curriculum. Whether it be in the classroom or at home at the dinner table, allocating time for students to ask questions and seek out new information can help nurture their natural curiosity and creative spirit.

Regardless of the specific methodology used, it is clear that in Israel and around the world education systems are keen to adopt approaches to learning that can boost creativity and cultivate innovation among young students. Traditional education systems are likely to continue relying on testing and rote learning. However, by recognizing the benefits of using these new approaches can positively influence the learning experience of developing minds both in China and across the globe.

# Chapter 5:
# The Refugee

The important thing is not to stop questioning. Curiosity has its own reason for existence. One cannot help but be in awe when he contemplates the mysteries of eternity, of life, of the marvelous structure of reality. It is enough if one tries merely to comprehend a little of this mystery each day.

-Albert Einstein

<u>My Father, the Refugee:</u> a personal story from the author
*Ami: The story of my father's life as a refugee is one that lives close to my heart. Since childhood, this story has and will continue to influence the way I see myself and the way I see the world. The story begins with my father as a young boy, growing up in a small village in Transylvania during World War II. He was six years old when suddenly he heard bombs begin to fall. Together with his mother and younger brother, my father grabbed a pillow in one hand, a prayer book in the other, and just started to run. Leaving everything behind, they ran east across the bridge outside their village and into the forest. By a stroke of good luck, the Germans decided to stop their siege at the bridge, giving my father and his family a chance to escape the immediate threat of capture by the Nazis. As they were running through the forest, together with the rest of the villagers who managed to make it across the bridge, they*

*miraculously found my grandfather who had arrived by horse. Together as a family, they began moving east towards Russia.*

*Over the course of the following two years, they would move from field to field hiding from the Nazi forces. It was during this time that my father lost both his younger brother and his father due to malnutrition and illness. They had no choice but to keep moving and before long my father had spent his whole childhood as a refugee wandering the frozen forests of Europe without food, without shelter and without any formal education. After the war ended, the family members that had managed to survive tried to find their relatives through acquaintances across Europe and the United States. This process was incredibly difficult, relying mostly on word of mouth and rumors to find the status of their extended family and close friends. When my father and my grandmother finally managed to make it back to their village, they found that the Nazis had killed everyone, including all of my father's extended family. This is when they decided that they would emigrate to Israel, an act that has a special word in Hebrew, "Aliyah", derived from the Hebrew word for "rise up".*

*Fast-forward a couple of decades and you'll find me, the youngest in my family, fascinated and mystified by my father's life as a refugee. I spent my entire childhood hearing about my father's miraculous survival and my grandmother's heroic actions that kept them alive. From almost ten years, all my father knew was running away into the forest, finding food and doing whatever it took to survive. From my perspective, these were more than just memories - these stories were the essence of who my father was. He often recalled the people who surprised him - Jews that were supposed to help but didn't, Nazis who were sympathetic to their situation and treated him and his family with kindness. I remember him telling me about the wounded German soldier they encountered in the forest that used his helmet to make them soup and shared his last meal with two Jewish kids he was sent to kill. Soon after, my father told me about the Jewish man who whipped him and his brother for begging for food as he rode past. Even years later, sitting in our home in the southern Israeli city of Ashkelon, it seemed that my father was both moved and perplexed by these unexpected, contrasting moments of compassion and humiliation.*

*As a child, I was perhaps most impressed with his stories of the seemingly superhuman powers that refugees like him and his mother used in order to survive. My father spoke of my grandmother the way you would talk about your biggest hero. Against all odd, she fought with every ounce of strength she had to keep her and my father alive. From running through forests in the freezing cold to eating stolen piles of potatoes for sustenance, she was not going to give up. Like many Holocaust survivors, for the rest of his life my father treated food with such reverence. He was the happiest person alive with a full plate of food in front of him – for him, this was pure joy. Despite us growing up in a household where, luckily, we never had to go hungry, my father's starvation-filled childhood never left his consciousness. One of my strongest memories of him is how he would keep always a few pieces of bread stashed away, together with other emergency supplies, ensuring that we were ready for the moment when we might have to flee. I grew up in a household where It was a crime to throw away even a piece of bread. To this day, I can't stand the sight of perfectly edible food thrown into the trash. I immediately think of the millions of starving refugees all over the world, searching desperately for their next meal. Following in my father's footsteps, I still leave our food scraps outside for the birds.*

*Long-term starvation, compounded by the constant threat of death, will change a person's psyche in a way that only refugees can truly understand. And yet, the collective memory of the Holocaust and our many years in exile as Jews is deeply embedded in the Jewish experience. It's part of our DNA as a people and it is certainly part of my DNA. For better or worse, Jewish culture, traditions and our drive to succeed is the result of the tens of hundreds of years we spent as refugees. It goes beyond the individual stories of people like my father – it is woven into the fabric of who we are as a people.*

*A big part of the refugee experience is the need to reinvent yourself. Some of the most amazing people that I have learned about in Jewish history – even Albert Einstein – arrived on foreign lands with nothing more than their motivation to succeed. When my father arrived in Israel he didn't speak any modern Hebrew and had no formal education. The only Hebrew he knew was from ancient Jewish prayers, but he didn't let this stop him. He would study day after day, devouring book after book to learn whatever he could. Years* later, I can say with the upmost certainty that his knowledge of history was better than any university graduate I've ever met. The voice in his head that kept saying, "I will survive, I can do it. Whatever life throws at me, I can handle" - that is what it means to be a refugee. This mentality is what enabled him to later become part of a legendary military unit, commanded by former Prime Minister of Israel, Ariel Sharon. Nothing would keep him down - not losing some of his closest friends in battle, not getting wounded twice during Israel's wars with neighboring countries. After what he went through as a child, it felt like nothing could stop him.

The undeniably strong and innovative spirit of refugees is inextricably tied to the horrible, unspeakable tragedies we have experienced as humankind. While Jews around the world often lead very different lives - religious and secular, living in Israel or part of the Diaspora, we are connected by this common thread of the refugee experience. The Jewish people have an innate understanding of what it means to start from nothing. For many of us, it means we have no choice but to be fearless, like my father. Despite the relative comfort and stability I've been privileged enough to have throughout my life, I was raised to believe that I could become a refugee at any time, for any reason. Much like my father's stories, this belief lives on in my heart and forces me to realize that the only things that truly belong to me are the skills and knowledge I acquire and my passion for innovation.

As an adult, I remain very sensitive to the issues faced by refugees around the world. I am particularly heartbroken by the stories of child refugees - victims of circumstance, just like my father, trying to survive. The global international refugee crisis we are facing feels incredibly personal. At times I cannot believe that, despite all the wealth and abundance in the world, there are so many with nothing. And yet, when I think about refugees like my father, I am inspired by the courage and fight with which they not only accept but overcome the many challenges they face. These are individuals that can truly change the world.

In order to survive and thrive in the future, we will need to adopt a mindset similar to that of Jewish refugees of the 20$^{th}$ century. This approach was not about being the smartest or the most education but finding a competitive edge. For many Jewish refugees this meant innovating within a range of scientific fields. The philosophy behind this trend was explored by Jewish American journalist and author, Thomas Friedman. Friedman writes, "the Twentieth Century began with massive migrations of Jews, to the United States, to the cities of Russia (and then the Soviet Union), and to Palestine. In each of these new lands, Jews turned to science in great numbers because it promised a way to transcend the old-world orders that had for so long excluded most Jews from power, wealth and society. Science, based as it is on values of universality, impartiality and meritocracy, appealed powerfully for Jews seeking to succeed in their new homes. It is not so much what Jews were (smart, bookish) that explains their success in science, as what we wanted to be (equal, accepted, esteemed), and in what sorts of places we wanted to live (liberal and meritocratic societies)." This search for a competitive edge is not limited to Jewish refugees. In fact, this same approach was adopted by Chinese immigrants in America, first during the gold rush of the 1850's and later in the establishment of the US transcontinental railroad.

It is estimated that between the beginning of the California gold rush in 1848 until approximately 1882, over 300,000 Chinese immigrants came to the US to try their luck at a creating better future for themselves and their families. Risking a grueling trip across the ocean and dangerous working conditions, these immigrants arrived in a foreign land in order to chase their dreams. Relatively quickly they noticed that the White and non-Asian gold miners tended to work individually or in small groups in hopes that they would find a gold payload that they could keep for themselves. They also found that those who worked on their own or in small groups were more susceptible to violence and discrimination. In making these observations, the Chinese workers realized that they had strength and power in numbers. They formed large, highly organized teams. Working as a unit, they split their profits. They also remained safe because they outnumbered the smaller groups of Gold miners who had previously posed a serious threat. Additionally, because they worked in larger groups, they were often able to cover more land and take in a bigger haul of gold. As such, even though they split the profits among a larger group of works, they were able to increase their individual profits.

When the dust of the gold mines cleared, and the great mining rush had all but been abandoned, many of the miners turned to working on the Transcontinental Railroad. The lack of relevant experience and the stereotype of being small and weak rendered most of the Chinese miners an unlikely choice to work on the Central Pacific Railroad. But once again, Chinese immigrants overcame this significant disadvantage by using their competitive edge. Rather than become discouraged by the fact that they were being paid less than White day laborers and being treated poorly, Chinese workers realized that they could apply their unique advantage to this scenario just as they had with the gold rush. Working in well-organized teams, they quickly became an invaluable resource for the railroad companies.

Accounts have it that after just one day on the rail line, the company decided to recruit and hire as many Chinese workers as possible. This included both those already located in California as well as workers in China who could be shipped to the American shores. In large teams, Chinese workers took care of each other and improved their collective wellbeing. Together, they were able to make gains in their fight for better standards of living, shorter work hours, and better wages. They used their competitive edge to make their way, step by step, towards a more prosperous life that would sustain their families and loved ones in China. Their innovative attitude and willingness to face the challenges of the unknown set an example for those who followed. They set the standard for what would later be recognized as the Chinese-American entrepreneurial spirit.

There is clear evidence that the Chinese people, particularly those who emigrated to foreign lands, have always been capable of adopting the refugee spirit and innovating their way to a better future. This heritage of innovation and risk-taking can be instrumental in our approach to cultivating these same characteristics among students in modern day China. However, in order to do so, we must recognize the limitations and problematic aspects of the current culture of education in China. While rote learning is highly regarded in Chinese education traditions, if it is not supplemented with opportunities for critical thinking, these ancient learning customs will remain a means to an end for knowledge acquisition. As a result, students are left with an elementary level of understanding of the core ideas they have studied and an inability to leverage their knowledge in the creative process. Repetition and memorization as the sole means of learning does not leave room for imagination and innovation. This issue is clearly demonstrated by Chinese and Chinese-American students in the US who have built an academic brand of diligence and impressive school rankings. And yet, following their studies, these same students are unable to translate these achievements into success as innovators and idea-makers.

Whether we are using examples from Jewish or Chinese history, we can see that when it is a matter of life or death, persevering in a new country or remaining empty-handed - uncertainty and instability are often the precursors to innovation. It is in such circumstances that people are forced to examine the world around them and adapt. In the case of the Jewish people, history has shown how being forced out of one's home, time and time again, has transformed the 'wandering Jew' into an intuitively innovative thinker. Now we must ask: How can we emulate aspects of this refugee experience within the Chinese education system in order to cultivate the next Nobel Laureates? How can we transform the educational experience of our students from an accumulation of information, test scores and class rankings to a more open, engaged and mindful learning process? Below are the central lessons we can take away from our understanding of the Jewish refugee mentality and practical applications of these lessons within our student's day-to-day life:

1) <u>Thrive from instability</u>: Much like Jewish refugees turned innovators, Chinese students much learn to embrace uncertainty and be willing to "throw out the rulebook." Thriving outside of your comfort zone - like a refugee, with nothing to lose – forces your mind to dig deeper and think critically. While this is a lifelong process, parents and educators can start with small but feasible challenges, such as navigating in a new neighborhood without a map or cooking a new dish without a recipe.

2) <u>Find your competitive edge</u>: Many Jewish Nobel Prize winners were forced to be refugees at some point in their lives – the most prominent examples being those affected by the catastrophic antisemitism of the Nazi party, the Fascist Italian governments, and the leaders of the former Soviet Union. When faced with the reality of losing their homes and their livelihoods, these Jewish refugees sought out how to use any advantage they could to get ahead and succeed in their new lives.

3) <u>Invest in education and the acquisition of new skills</u>: As a result of their nomadic existence, the Jewish people quickly realized that the only thing it made sense to invest in was their education, literacy and ability to learn new skills. Not knowing when they would next be expelled from their homes or when their material goods might be confiscated, Jews poured their time and resources into their minds and those of their children. While Chinese parents are traditionally very committed to the education of their children, they focus this commitment on ensuring their students achieve academic excellence.

Ask yourself, what would my student need to know if they were left with nothing? How would he or she be able to use their knowledge and skills in order to create a new life for themselves? These are the skills and type of education pursued by Jewish refugees.

4) <u>Continue to learn and discover new information</u> – Students may assume that receiving marks of 100% and excelling on tests means they have all the information they need. Part of the refugee spirit is questioning the information presented and seeking out new perspectives. How does your student respond to information that contradicts what they may have learned? Aspire to improve their willingness to reexamine existing knowledge, adapt based on what they've learned and seek out new solutions. In doing so, parents and educators can demonstrate to their students the way in which refugees are forced to react to dynamic, complex challenges. This new information and discovery process can mean the difference between being awarded a perfect GPA and being awarded a Nobel Prize.

# Chapter 6: Language

The greater part of our knowledge and beliefs has been communicated to us by other people through the medium of a language, which others have created. Without language our mental capacities would be poor indeed, comparable to those of the higher animals.

-Albert Einstein

<u>Language as a Passport to the World:</u> a personal story from the author
*Jordan: I grew up in a small town in the southeastern part of China. My mother tongue is a southern dialect spoken by only 300 thousand people. People from other cities would frequently have a hard time understanding us. Although we spoke a dialect at home and among our family and friends, Mandarin is the official language of China and it is the language used in school. While I know I must have learned Mandarin at some point in early childhood, I don't remember when or how it became my second language. Language acquisition has long been a part of the Chinese experience, though this experience evolved over the years. My mother learned Russian when she was in school as part of China's efforts to model itself after Russia at the time, including the Russian education system. My mother must have found this experience influential because in the 1980s my mother decided to teach my sisters and I English. I'm not sure where she got the motivation to do so as this*

*was long before English was offered in our schools and it was only a few years after Sino-US relations had resumed. Nevertheless, she was determined to add English fluency to our skill-set.*

*The process began with my mother first teaching herself English words and sentences and then she would slowly pass on to us what she had learned. We began with simple concepts and vocabulary words: numbers, colors, shapes, foods, etc. As we progressed, my mother began to teach us short pieces of dialogue from a variety of plays, giving us an opportunity to act out the words and gain confidence and a sense of fluency in this new language. What started as my sisters and I playing at home and learning how to interact in English using dialogue transformed into a talent showcase of sorts that we would put for friends and guests in our local community. We learned the English language through play, performance, and friendly competition. My childhood memories are filled with fun and laughter and I was blessed to learn in a way that motivated me rather than pressured to succeed. While I actually didn't truly get the opportunity to practice my English until college, there is no doubt that these early interactions in English were critical to my ability to excel in the language later in life. They also made me just a bit less nervous about saying "hi" to a foreign student on campus.*

*Despite my multilingual upbringing, it took me a long time to truly appreciate the beauty and benefit of understanding another language and the true essence of communication. Being able to speak a language isn't only about communicating efficiently with others. It is the basis for building meaningful relationships with those around us. When I achieved true fluency, I found that I was able to understand the meaning behind the words and could use language to create connections. Language and being multilingual not only helped me find my way in the world but made the interactions along the way that much more meaningful. It helped me bridge gaps between seemingly opposing ideas, cultures and perspectives. I also became very aware of the language abilities of those around me and how I could learn from their adept use of language. The conversationalists I most admired were those who used their language skills to go beyond small talk and were able to make the person sitting across from them feel at ease and understood. Decades after those short English plays I put on with my sister, I am a staunch proponent of advanced language acquisition and multi-lingual studies. Whatever the goal we set for ourselves, I believe that language is one of the most powerful tools we can use to succeed.*

Language comprehension is one of the most challenging tasks for our brains. Perhaps that is why when we are able to comprehend two or more languages, our brains are able to make new connections and improve not only our understanding of each language but of the world around us as a whole. Not only can a second language be the key to traveling around the world and exploring a new culture, it has been proven that the acquisition of second language can expand and improve overall brain function. In one study, Swedish military recruits were taught a new language. Brain scans before and after their training depicted an increase in the size of the hippocampus, the brain's language center, immediately after the language session. In another study, language acquisition led to an increase in the number of neural pathways between different parts of the brain. Similarly, in a study conducted wherein English speakers were monitored while they learned Chinese vocabulary, MRI results showed that not only did the language learners develop better cross-regional brain connections, their brains showed structural changes that were present even six weeks after completing the learning process. This proves that language can play a critical role in enabling our brains to grow and change over time. When exploring the effects of multilingual capabilities on non-language skills, studies have shown that individuals who speak two or more languages have better cognitive abilities than those with fluency in a single language. Likewise, research shows that acquiring a second language can improve everything from decision-making skills to an improved ability to understand one's surroundings. Language acquisition can lead to increased cultural understanding, an ability to see the nuances of differing perspectives and the ability create a bridge between cultures. All of these capabilities are critical precursors to one's ability to create and innovate.

Most Jews, whether they grow up in Israel, the Diaspora of Europe, or even South America, are multilingual. Just as Chinese citizens often have a local dialect, learn the official language of the province, and study Mandarin and English in school, so too do Jewish people around the world have a connection to the spoken and written word in multiple languages. The modern language of Israel is spoken Hebrew, but prayers are sung in ancient Hebrew which has origins in Aramaic. While the modern Hebrew language and the Hebrew found in religious texts are connected through the three-letter roots that serve as the basis for almost all modern Hebrew vocabulary, these are in many ways two distinct languages that are learned completely separately. In some cases, regional languages were formed, such as Ladino, a Judeo-Spanish language developed around the Middle Ages by Jews of Spanish origin. Similarly, Jews born in the diaspora of Europe and Asia were most likely to grow up speaking Yiddish - a mix of old German and ancient Hebrew. However, in order to interact with the non-Jewish communities around them they would also learn their native country's language, such as Russian, German or Italian.

In each instance, Jews had the language they spoke within their homes or in their local villages, another language they used for prayer and religious customs, and a third language of their native country that would enable them to interact with others for purposes of trade, commerce and education. It was commonplace that no matter where Jews met around the world, they would always be able to communicate in one of their shared languages, be it Yiddish, Ladino, or Hebrew. This meant that the majority of the Jewish population, although scattered throughout the world- had a uniting characteristic available to them in order to communicate. This historical and cultural tradition of being multilingual helped Jews across the world succeed in everything from escaping to a new country as a refugee to exploring scientific collaborations that would one day lead to receiving a Nobel Prize. Through their efforts to become multilingual, the Jewish people were and continue to be able to communicate effectively, giving them an innovative edge in our increasingly global and interconnected society.

Perhaps one of the more interesting stories of Jewish innovation that was enabled by being multilingual is that of the Rothschild family and their famous banking empire. What today represents a global force in everything from business to politics to culture, began with the birth of a Jew named Mayer Amschel Rothschild in the Jewish ghetto of Frankfurt am Main, Germany. Despite his later success, Mayer came from very humble beginnings. His father was a simple tradesman who made a decent living for the family collecting and trading coins for the Prince of Hesse. Mayer grew up in a house with more than 30 family members, living right above his father's small shop. Although Mayer was supposed to devote his life to Jewish studies and began his rabbinical studies in the German city of Furth, at the age of 12 his life took an unexpected turn.

With the sudden death of his parents, Mayer instead left school for Hannover, where he became an apprentice at the firm of Wolf Jakob Oppenheimer. Mayer worked his way up in the financial company that dealt in international trade. He would soon marry and in the following years his wife would give birth to five sons and five daughters. From the moment his boys were ready to be apprenticed, he began cultivating within them a worldview that would eventually turn Rothschild into a household name for the remainder of his lifetime and for generations to come. At the turn of the nineteenth century, Mayer Rothschild sent his sons to the five business capitals of the world at the time - Frankfurt, Naples, Vienna, Paris, and London. His vision was to have each of them establish a bank in the Rothschild name that would enable them to become a global banking operation. Their success was revolutionary and remarkable. By leveraging their family business at each of the major financial centers of the time, they were able to expand and grow year after year, eventually becoming the Rothschild banking dynasty we know today.

How did the sons and Mayer turn from new immigrants to banking leaders in each of these new cities? Language. Mayer Rothschild recognized that in order to conquer the banking world he and his sons needed to be fluent in the language and culture of the cities where they sought to establish business operations. The Rothschild sons spent a great deal of time and energy gaining the fluency in each of these new European languages such they could act as a local rather than as a stranger within their new environments. Were it not for learning these second and third languages fluently, the Rothschild brothers would have been much less effective in their international banking efforts. As a result of their fluency, they were able to work in close collaboration within the local markets of each city. Needless to say, this approach paid off. The Rothschild banking empire went on to influence over a century of business and culture in most of the major European cities as well as in the UK and the United States. The success of the Rothschild family is still referenced in their current website prospectus, describing the innovation and success of their family as such: "At many times during our 200-year business history we have had the boldness to hold a distinct perspective and take an innovative approach. This heritage provides us with both a strong foundation and a vision for the future." Fluency in a foreign language was a key factor in realizing the Rothschild's vision and it served as a building block for what remains an influential business and investment group today.

Given the international and global nature of our current society, it can be argued that we are on the cusp of the adoption of one overarching language that will be required in order to communicate globally. While some may think the most natural answer is to expand an existing language already spoken among many millions of people, such as English or Chinese. However, there is already an additional language that has been integrated into most of society and it is the most common language on planet earth. We are referring to the language that Ami defined in his TED Talk as Robotish. What is Robotish? While you may have never heard the specific term for it, you are already very familiar with it. If you own and operate a computer, if you surf the internet, play games or download apps - you can 'speak' Robotish. While almost everyone in the world can 'speak' Robotish, few are actually able to read and write the language. Robotish is in essence, the language of code and is used for all aspects of computer programming. Everything that requires computational function, from central heating systems to online learning programs is based on the use of Robotish. Coding and software must be able to be created, read, and 'spoken' as the next step to creating enterprises as successful as that of the Rothschild family. Robotish trains young minds to both think outside the box and at the same time gives immediate feedback, enabling and cultivating the creativity of each coder. By leveraging this inherent 'feedback loop,' Robotish could become one of the easiest languages to add to our skill-set and it merely requires that we all agree to invest the time to learn. More importantly, as we have seen over the past two decades, those who are able to integrate this language into their lives as students and leverage it as professionals are likely to lead the charge in developing the technologies that will shape our futures.

As with other examples we have provided throughout this book, the Jewish people – and in this case, the people of Israel – are leading the pack when it comes to cultivating innovation through the acquisition a critical skill. In this case, it's Robotish. In fact, in Israel coding has actually been a mandatory high school class subject since 1996, making it the first country to establish such parameters and guidelines. There are already 'coding kindergartens' in Israel as well as numerous programs aimed at expanding coding education within elementary school curriculum. The State of Israel recently announced that it will be forming national cyber education centers in order to train the educators who will be responsible for teaching Robotish to students in schools throughout the country. The Israeli high school curriculum for studying Robotish currently includes three main elements. First, students must develop their fluency in computer languages in order to understand how to write code using the algorithmic thought processes. Next, they must understand the structure of computers and the internet as a whole. Lastly, they must develop the ability to analyze existing computerized systems and cultivate their ability to think creatively within the framework of such systems. The methods of instruction that are already in use in Israel are based on active learning environments wherein students meet two to three times weekly and complete hands-on challenges to ensure they not only understand the nature of Robotish as a language but are able to put it into practice. This national program in Israel is supported by the Ministry of Education and is implemented with the cooperation of the IDF. Given the role technology, innovation and problem-solving play in the success of the Israeli military, it is no surprise that the IDF is involved in the advancement of such programs.

Everyone from educators to parents to military leaders in Israel understand that learning Robotish will not only give students a competitive advantage as adults, but it can also provide them with opportunities to develop their ideas. The language of Robotish teaches students to break problems into workable pieces. While non-programmers might find this approach somewhat foreign, coders fluent in Robotish will naturally begin to problem-solve by breaking larger challenges into small, manageable obstacles. They begin to think more logically, and therefore, they begin to find creative solutions while working their way through a structured format. Students can learn to excel in this new skill and use it to create opportunities for themselves as they pursue higher education. While kids in Israel might begin using Robotish to deepen their understanding of existing school curriculum or explore their creativity by developing computer games, this is just the beginning. As students' fluency in Robotish improves, they will be able to use this same language to develop the solutions for challenges we will face as a society in decades to come.

In both the previous chapter and those to come, we seek to translate a particular element of the Jewish cultural infrastructure for innovation into a variety of lessons and practical applications for Chinese students – here our suggestion is both singular and simple: <u>learn and become fluent in Robotish</u>. Chinese students, particularly those seeking a competitive edge over their peers, have an enormous amount to gain by learning all aspects of Robotish. Beyond the more obvious computer and coding skills that students will learn through their fluency in Robotish, they will also learn to expand their methods of thinking and processing. Learning Robotish may be nothing less than life-changing for Chinese students. The world is integrating this new global language at an unprecedented pace. We will soon live in a world where every interaction with the world around us will be built upon a computer platform. By prioritizing the acquisition of Robotish today, students in China can stay ahead of the curve and go on to lead the world as the innovative coders of tomorrow.

# Chapter 7:
# "Balagan" - Chaos

The eternally incomprehensible thing about the world is its comprehensibility.

*-Albert Einstein*

*Disrupting a Chaotic Industry:* a personal story from the author
*Ami: Early in my entrepreneurial career, I founded a 3D TV technology company that had a particularly groundbreaking solution. As we began growing the company and scaling up our offering, I realized just how chaotic this industry was. We knew that there would be able to create an amazing visual experience that would hook new clients. At the same time, the cinema and film entertainment industry seemed satisfied with the status quo of using 70-year-old, 3D technology. Despite knowing we had something great at our fingertips, we know Hollywood was overwhelmed by the process of converting to newer digital media, and we weren't sure if the movie studios would be ready and able to adopt our disruptive technology. Would they have the budget to make one 3D movie a year? What about twenty? Despite our deep understanding of existing industry standards, it was extremely difficult, if not impossible, to understand the kind of pipeline our technology would travel through to reach in order to reach the end-consumer. How many of the 3D movies produced*

for the big screen would even end up reaching people's homes? There were so many unknowns.

The first thing I thought was, "what a balagan!" Balagan, the Hebrew word for a chaotic mess, was the best way to describe the challenge we faced as a company. In addition to us not being able to predict the most suitable application of our technology, we found that every country had their own standards about the introduction of 3D technology. Some countries, France for example, loved it and thought it would usher in a new era of cinema. Other countries like Italy thought it would cause health issues such as eye infections and moved to block the development and adoption of all types of 3D technology. The lack of a consensus extended even to countries that all viewed 3D technologies in a positive light. There were the nations that wanted to integrate what was called 'active 3D' while other nations preferred 'passive 3D' technology. In short, there was nothing even close to an international standard that we could plan around.

With the mayhem and balagan around us, we understood that if we wanted to be successful we needed to move fast. If we could enter the market quickly with one 3D solution, we could jump out ahead of the industry. For our first attempt, we used 6 projectors to create a 3D solution. After a short while we realized that this would definitely not be the final product and moved on to embrace another technology. Likewise, while we initially tried to market 3D technology as a cinema attraction, early on we realized that there just wasn't an expansive enough 3D movie database. In order to bolster this offering, we included our own library of classic films. As quickly as I could, I ran with this new idea. And this is how we evolved. Each time we tried to find a new technological direction or new niche in the industry, we kept one thing in mind - speed. If what we had developed was successful - great! If it failed, we pivoted and took off running in a new direction.

In some ways the fast-paced, fail-forward approach that I had as an entrepreneur was first developed during my days in the Israeli Secret Service. I remember one of the special surprise military training exercises that we were given wherein we were dropped in an unknown location and told to "figure it out." There is a door in front of you and once you open it, you're thrust into a *completely unexpected environment full of fellow soldiers, enemies, victims, etc. You don't have time to start analyzing or trying to understand the situation. The purpose of these surprise drills is to force you to act quickly in an emergency situation where all your emotions are heightened. The only thing you can do to succeed in the exercise is to move as fast as you can in the direction that you believe is the right one. If you fail, you choose another path and run just as hard in that new direction. The most important thing is to never stop until you successfully completed the drill.*

*That was the only way to approach these military exercises and it became the only way that I approach my business ventures. It is in these complicated situations where everything seems out of order – a total balagan – that you must pursue your goal while moving full speed ahead. It doesn't matter if you're working with a new solution, theory or product, you must be able to test extremely quickly, regardless of the outcome. This approach is at the core of Israeli culture. We test quickly and if we fail, it's just the next step in our process. Instead of slowing down into failure, you must accelerate, thereby giving yourself the momentum and motivation to find your way through the inevitable balagan you will encounter. Sometimes you'll find the best path on the first try; sometimes it takes a few wrong turns before you get there. But if you're relentless in your pursuit and move as fast as you can – you will be able to do what others can't – get there first!*

The word *balagan* is used daily by Jews to describe anything from their children's rooms and daily traffic to political disarray and stock market fluctuations. The connotation of the word *balagan* is rarely positive and often involves disciplining a child for some sort unruly behavior. But the word is not specifically unkind either and is considered an inevitable part of daily life. This word is often used when describing one of the most core elements in Jewish life that contributes to their cultural infrastructure of innovation – education.

Israelis often refer to the education system as a complete *balagan*. This small nation's budget is approximately US$ 12.5B Billion and in 2011 the country spent 7.3% of its GDP on all levels of education. While it is far from perfect, education in Israel is considered priceless and teachers receive an adequate salary as part of a strong funding program. Historically, international leaders from all parts of the world have praised the Israeli educational system for producing such an impressive body of successful students that have gone on to with acquire advanced graduate degrees.

That said, parents from outside Israel would be stunned if they walked into a 1st grade classroom. From the busy environment to high noise level, children addressing teachers by their first names to not sitting in assigned seats, the Israeli education system is extremely informal. Students often drag chairs around the classroom during the day to socialize with particular students they are friends with. In a room of 35-40 six-year-olds, the teacher somehow is able to manage getting his or her point across. It doesn't take long before the classroom turns into a platform for negotiation. When it's time for recess and the teacher says they'll have 15 minutes outside, students immediately start trying to get themselves a better "deal." One student might ask, "can we have 20 minutes instead because we were so well-behaved this morning?" Once that interaction is complete, the teacher will let the students know that they can bring only five toys outside with them. The students immediately begin to negotiate in small groups about how best to split up the toys for their recess and eventually make the case to their teacher. A teacher's statement is most often met with an alternative solution. Classroom rules were made to be broken, or at the very least, questioned.

As you can imagine, with so many questions, negotiations and independent thinking, the classroom can get noisy and rambunctious. It is no surprise then that these 1st graders become familiar with what it means to deal with this *balagan* at a young age. They are forced to think creatively, both as individuals and in small groups, in order to negotiate and get what they want. Teachers accept and encourage this behavior by addressing questions quickly and not letting classroom discussions get out of hand. They do not chastise the child for asking and understand that doing so is part of the learning process. In the Israeli classroom, asking questions is part of internalizing facts, making educated guesses and developing creativity. What might appear as chaos might in fact be an important part of how these young students create order in their minds. Inbal Arieli, serial entrepreneur and senior advisor to the authors of "Startup Nation", was able to very accurately describe how the concept of *balaga*n can promote creativity and independence: "*Balagan* goes hand-in-hand with freedom of thought and creativity. As anticipated, recent experiments into the effects of disorderly environments on behavior concluded that while orderly environments encourage conventional behavior, messy ones stimulate a generation of new insights".

When a Chinese parent sees a playground full of children climbing up the slides instead of sliding down, standing up on swings instead of sitting, and kids moving constantly in the direction of their next adventure – their first thought is probably: what is this chaos and how can we stop it?! However, Israeli entrepreneurs like Arieli see the same scene and think – this is great! It is the *balagan* that teaches children social order. Parents in Israel will rarely interfere in the children's playing. They allow the social experiment to take place as long as no one is in imminent danger of physical harm. *Balagan* teaches kids that there isn't always a 'right' way or a 'right' order in which to do things. This unstructured play is both socially and intellectually challenging and can be critical to developing the minds of young, innovative thinkers and risk-takers. Arieli continues: "*Balagan* can invite conflict and frustration but it also demands on-the-spot solutions for specific situations and people, as opposed to relying on an all-purpose rule supposedly fit for any circumstance. So, the mechanism of disorder in fact creates order." Because it is so built-in to

Israeli society, the *balagan* that exists in the classroom or on the playground continues on to be part of the way adults in Israel operate in their day to day lives. It is not limited to those who are naturally more prone to leading chaotic lifestyle, such as painters or musicians. *Balagan* is a tool for thinking that is relevant for everyone from lawyers to doctors, managers to scientists. By relinquishing the need for order, Israelis leave more room for nurturing a creative spark. For better or worse, the lack of order elevates inspiration and leaves an open page for whatever might come next.

Seeing the benefits in chaos in the school system and the resulting creativity is not limited to the Israeli culture and Israel's education system. In one of the most-watched TED talks of 2006, Sir Ken Robinson asked, "Do Schools Kill Creativity?" Robinson, an education thought leader argued that creativity is "as important in education as literacy, and we should treat it with the same status". On a more philosophical level, Swiss psychiatrist and analytical psychologist Carl Jung famously wrote, "in all chaos there is a cosmos, in all disorder a secret order". Although the most natural first response to chaos should be discipline, by enabling and encouraging a certain level of disorder, we can ignite individuality and innovation. While many aspects of Chinese traditions focus on order and control, Chinese culture also has its own insights and references to the importance of *balagan*. For example, the Chinese word for crisis is defined as both danger and opportunity. In other words, a crisis, or chaotic scenario, inherently includes both the risk-taking nature (danger) cultivated by such environments, and the creativity and innovation (opportunity) that students can explore when they are given the chance.

Another example of the value attributed to *balagan* in Chinese culture is the way traditional Chinese medicine approaches the healing of physical illnesses. Chinese doctors don't see symptoms of illness as chaos that must be stopped, but rather finds a way to work within the chaos in order to bring the body back to good health. While Western medicine often seeks to rid the body of the symptoms of illness, Chinese medicine understands that this chaos plays a role in the body's healing process. Using the circular ebb and flow of energy within the body, Chinese healing practices acknowledge that the chaos of illness is not something to be feared but rather an opportunity for the body to seek out what it needs.

Through our examination of the concept of *balagan*, we have created a clearer picture of how an Israeli classroom compares to a Chinese classroom. We can see how the chaos Israeli children encounter and the elements of discovery and play may enrich their innovative abilities long-term. By gaining a greater understanding of the role chaos, or *balagan*, plays in Israel's cultural infrastructure of innovation, we can seek out the ways in which Chinese parents and educators can integrate *balagan* into the lives of their students. We fully understand that integrating an acceptance of chaos into a Chinese student's academic life won't be without its challenges. However, the first essential step is for parents and educators to acknowledge that the disorder, the unknown, and the unexplored play critical roles in cultivating innovative thinkers. Doing so does not mean disregarding the social and moral norms of Chinese culture or negating the importance of seeking order in one's life. It simply leaves space for creating a bit of the *balagan* Israeli children have in their daily life and reaping the benefits of this approach. Below are some of the most important lessons we can learn from *balagan* as well as practical applications that can be of great use for Chinese parents and educators.

1) <u>Start from day one</u> - All we have to do is look at a newborn baby to remember that a sense of creativity and discovery are in our DNA. Babies spend their first few months of life figuring out how to roll over, wave, crawl and then walk. During this time, they innately experiment and try until they find their way. As children, we create games out of everyday items, often imagining fantastical worlds and our desire to explore them. Chinese parents and grandparents can

give this ability back to their children and help Chinese students remain as innovative and creative as they were in their first years of life. *Balagan* and disordered environments tap into this creativity early on and encourage children to maintain it throughout their education and into adulthood. Rather than engraining in students the logic that there is only one correct answer or solution, Chinese parents and educators can create a *balagan*-like environment by encouraging their students to be proactive in finding the answers in their own way. This may mean ignoring what is considered the "right" way and instead focusing on the creative process used to reach a solution. The creative thinking that comes from *balagan* is like a muscle and it is a skill that must be given ample opportunities for growth. By using this skill continuously throughout their childhood, Chinese students can eventually leverage these same abilities as adults in order to innovate the world around them.

2) <u>Find flexibility within the existing structure</u> – Striving to integrate a bit of *balagan* does not mean that all sense of order has to fly out the window. The consistent schedule within which each student operates is important and it gives them a sense of structure and security. However, it is within this framework that we would encourage parents and teachers to find opportunities for students to explore without boundaries. Perhaps try turning a piano lesson rooted in classical theory into an opportunity for a student to create his or her own composition. If possible, take time away from difficult subjects in order to study something new, even if it's only for a short time. A break from math to explore some poetry or foregoing ten extra minutes of biology in order to complete a small assignment about a different culture or language are just a couple examples of how this can be achieved. This flexibility and fresh perspective can not only break up the monotony of a daily schedule but also instill a sense of adventure in a student's learning experience.

3) <u>Let the *balagan* shine through</u> – Whether we like it or not, the world is a fairly chaotic place. New, unexpected developments happen all the time and they provide important opportunities to learn and grow. By accepting the underlying duality of *balagan*, Chinese parents and educators can move beyond teaching students that they must seek order in order to manage chaos successfully and instead explore the value of facing the chaos head-on. Rather than avoiding feelings of danger and fear by focusing on controlling the scenario, Chinese students can seek out the opportunity for innovation hiding inside a complex, messy situation. In acknowledging the value of this hidden chaos, Chinese parents and educators are allowing some of that *balagan* to permeate through into everyday situations. This will translate into give new creative thinking opportunities for students and an ability to see chaos as something positive rather than something to be feared and avoided.

4) <u>Asking "Why" and "What if"</u> - The most important questions that can be used to integrate *balagan* into daily life are 'why' and 'what if'. Asking these questions doesn't mean that a student does not accept proven facts, but rather that they are thirsting for more knowledge and a deeper understanding of any given subject. Get your student started on this path of questioning and inquisitiveness by reframing a subject or idea in a way that leaves room for curiosity. Much like a second language, ongoing practice is needed in order to improve the 'fluency' with which students ask these questions and seek out a bit of *balagan* in their knowledge acquisition process. Over time students will gather new tools and tactics for exploring ideas and concepts beyond what they are initially taught in the classroom or at home. By asking 'why' and 'what if' on an ongoing basis, they will be able to make new, cross-disciplinary connections and leverage this knowledge and creativity in order to innovate.

5) <u>Explore New Ways to Problem-Solve</u> – Innovators understand that the *balagan* is part of their process will often try completely unconventional ways to reach their goals. From turning their problem upside-down, sideways or taking it completely out of context – innovators are willing to do whatever it takes to stimulate their mind and find new pathways that can ultimately lead to success. Understanding that there might be many ways to reach a goal is best learned through practice. Students should identify their goal and try working backwards from the end to figure out where to go next. Make sure they check their progress along the way and are not timid about reevaluating if the steps don't get them where they need to go. These exercises are not about solving a problem as much as they are about willingness to try a new approach. Much like the other recommended lessons and exercises, putting these methods into practice on a regular basis will encourage students to explore a diverse range of potential solutions to the many challenges they face.

# Chapter 8: Debate

If A is success in life, then A = x + y + z. Work is x, play is y and z is keeping your mouth shut.

-Albert Einstein

<u>How Debate Restored My Self-Confidence</u>:
<u>a personal story from the author</u>

*Jordan:* People often describe me as fearless. They use words like bold, action-taker, and risk-taker. It's true- I don't have much fear in me. This is probably due to the freedom I had as a child. I was the third child in my family, a lucky survivor of China's One-Child Policy. I was sent to the countryside to live with my grandparents from age two to seven. I remember running and climbing mountains, swimming in the streams, catching fish and shrimp. I had very few limitations and my grandparents encouraged my adventurous spirit. I was one of the best students in my middle school. That didn't last long after I was admitted to Peking University. I was confronted with the best and brightest minds from around the country. Now faced with the educational standards of the "Harvard of China", I was in shock and awe. For someone who had always been described as daring, I had now completely lost my confidence. As a public speaking champion and student council leader in middle and high school, I had gained praise for my "impeccable" English by our small-town standards. But when compared with students from big cities, I found that I was far from outstanding.

My first year brought with it a series of failures: I didn't get to speak on the radio station due to my southern Chinese accent; I didn't pass my CET-4 English test; I wasn't picked as a student member at a student association; I wasn't accepted into the school dance troupe. I was deeply disappointed in myself. The only positive recognition I received was for my Indonesian pronunciation and how well I got along with professors, department leaders and older students. Even though I was awarded various honors and scholarships within my department, I hadn't been able to rebuild my confidence. There is a Chinese saying: 强中自有强中手, which translates to: "however strong you are, there is always someone stronger." At the time I wasn't able to realize that having strong competitors would raise my own standard of excellence. Instead I was focused on my deep-seated feelings of defeat.

My sophomore year brought with it a new opportunity that completely altered both my sense of self and my future. My department recommended me to run for one of the five seats on the Student Council Board of Directors. It was quite a daunting election process. Candidates needed to publicly advocate for themselves with posters and running slogans in the busiest areas on campus. I likened it to the Presidential election in the U.S. — all candidates needed to give a speech and take questions from about 800 student representatives from different departments and schools. Even though I felt uncomfortable, I knew that I just needed to get behind that podium for the magic and perseverance of my youth to kick in and inspire me. I knew how to debate. I knew how to think quickly on my feet. I knew that I must prepare and study in order to find a way to outsmart my competition and decisively win the student body vote. As I walked up to the podium that day, I was not thinking about winning or losing. I was focused on facing this election, just as I had faced every debate, public speaking lecture, and presentation – and in that moment, I was able to once again leverage my ability to think quickly on my feet.

*Shortly after the debate, I was elected as the only female vice president in that year's cabinet and I was put in charge of the students' activities at the university's Centennial Anniversary celebration.*

Educators across the globe will agree - debate creates future leaders. It gives students skills that will be critical to their future success, regardless of the profession they pursue. Beyond serving as a marketable and universally valuable skill, the ability to debate enables those that can think creatively to pursue their innovative ideas and become the "early adaptors" of their time. Given the link between debate and innovation, it is not surprising to learn that the origins of Jewish debate began over two thousand years ago. Abraham, one of the central figures of the Bible, venerated in Jewish, Christian, and in Islamic texts, was the first to question - or debate - God. As Genesis relays, God had decided to destroy the citizens of Sodom and Gomorrah because of the wickedness raging through the cities. Then Abraham approached God and said:

> *"Will you sweep away the righteous with the wicked?*
> *What if there are fifty righteous people in the city? Will you really sweep it away and not spare the place for the sake of the fifty righteous people in it? Far be it from you to do such a thing—to kill the righteous with the wicked, treating the righteous and the wicked alike. Far be it from you! Will not the Judge of all the earth do right?"*
> *The Lord said, "If I find fifty righteous people in the city of Sodom, I will spare the whole place for their sake."*
> *Then Abraham spoke up again: "Now that I have been so bold as to speak to the Lord, though I am nothing but dust and ashes, what if the number of the righteous is five less than fifty? Will you destroy the whole city for lack of five people?"*
> *"If I find forty-five there," he said, "I will not destroy it."*
> *Once again Abraham spoke to God, "What if only forty are found there?"*
> *God replied, "For the sake of forty, I will not do it."*

> *Then Abraham said, "May the Lord not be angry, but let me speak. What if only thirty can be found there?"*
> *God answered, "I will not do it if I find thirty there."*
> *Abraham said, "Now that I have been so bold as to speak to the Lord, what if only twenty can be found there?"*
> *God said, "For the sake of twenty, I will not destroy it."*
> *Then Abraham said, "May the Lord not be angry, but let me speak just once more. What if only ten can be found there?"*
> *God answered, "For the sake of ten, I will not destroy it."*

In this exchange Abraham successfully debated God for the first time - always questioning, never afraid to ask, never afraid to express his beliefs. Jewish religious history has a great respect for debate. It can be argued that debate is the central aspect of the expression of the Jewish religion. From Abraham to Moses to King David - the Old Testament is filled with examples of lively debate. But some of the liveliest debates in Judaism are those that took place between the Rabbis. Rabbis are the head teachers of the Jewish religion and are both highly respected and highly educated, having a deep understanding of the ancient texts as well as modern teachings. The most well-renowned rabbis of Judaism are those that created a written documentation of oral histories, teachings, and interpretations of God's word and laws, known within Judaism as the Talmud and the Midrashim.

Even the manner in which religious texts were annotated in ancient times reflects the important role debate plays in Judaism and the way in which Jewish scholars learn. The central argument from the Old Testament or Talmud is found at the center of the page. Surrounding this argument would be all of the notes and interpretations from Rabbis and Scholars throughout the years. This same method is still used today. None of the interpretations is considered the most important or the most popular, rather, each idea is given equal time to be analyzed. This principle can be found in many applications of Jewish debate – no one argument is considered strongest or weakest and all arguments deserve to be heard.

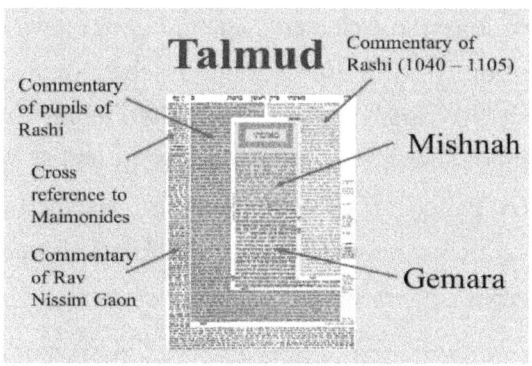

*Example of Text from Talmud and Surrounding Commentary*

As an exercise in the times of the great Jewish scholars, two opposing schools of thought, each following a particular set of religious interpretations and arguments that had been developed over time by venerated scholars, would spar over meaning and application of a tenet of Jewish ideals, beliefs or traditions. These details could be as small as a blade of grass and as large as the interpretation of the word of God.

During the times of the great Rabbis and Scholars of Talmudic debate, there were two very famous schools of study and interpretation: Beit Hillel and Beit Shammai (the House of Hillel and the House of Shammai). Jewish ancestors rallied around these two seemingly opposing houses and views of the interpretation of Jewish law. However, they treated each other with the upmost respect and continued the tradition of reaching communal decisions. Each house understood that the goal of the debate was not to win or lose, but rather to understand the value of each argument and how the debate process could shed new light on any given issue. Each argument was seen as a critical step on the path to finding truth in knowledge. When one describes the idea of debate, the image conjured is often of two swords sparring. From the perspective of Jewish scholars, instead of the swords breaking or shattering, debate enables each side to sharpen its sword by clashing against their opponent. This imagery can also be found in Buddhist and Eastern traditions, often referred to as the duel of ancient wisdom.

One of the most important elements of traditional Jewish debate is the role of collaboration. In order to conduct a debate, Jews require what is called in Hebrew "havruta." Coming from the Hebrew word for friend ("haver"), "havruta" describes the unique way in which ancient scholars and students were paired off to debate Jewish texts such that they could understand the entire argument. Jewish scholars believed that the best way to gain a deeper understanding of a text was to read it out loud with a partner. From there, the debate and questioning could begin. Havruta is usually done in pairs but can be expanded to several students presenting each side, similar to study groups. The process would be completed in the following manner:

1. The partners or small groups take turns reading a text out loud.
2. The students read the text for the second time, each partner or group reading alternating paragraphs.
3. Next, each student shares a short initial reaction to the text.
4. Once all students have offered up a short summary, they can begin to interpret the text more broadly or debate specific points of contention that they may have.
5. As they continue to exchange ideas, students encourage each other to think creatively, and have the freedom to express any argument without judgement. Their creative thinking is stimulated by their partner(s), who can both agree with their stance or play devil's advocate in response to the insights and interpretations discussed.

Unlike traditional ways of learning that are led by a singular teacher, the practice of havruta puts everyone on an even level from which to dive into the discussion. A learning unit of two or more people creates a dynamic where each individual can act as both teacher and student. Likewise, there are no right or wrong answers or limitations on the exploration process. Additionally, traditional havruta required a complete mastery of a particular text or concept. A student of one of the

Rabbinical Houses will always preface the argument with Rabbi X states this, and therefore argument XY is valid. Rabbi A states this and therefore argument AB is valid. Preparation for a havruta required a meticulous review of the material and the rabbis who excelled in this format were capable of referencing a vast array of points made by the scholars who came before him.

Fast forward to today, there is an engrained sense of debate and the need for questioning within Jewish and Israeli culture. From childhood, Jews are encouraged to seek answers and debate one another in order to better understand their daily lives. By continually asking questions that others have not yet asked, individuals are invariably able to arrive at new ideas and innovative solutions. Despite how it may appear to outside cultures, Jews and Israelis are not just debating for the sake of arguing. They debate to truly see the argument from all sides. Sometimes there is no right and wrong answer, but rather additional layers to be revealed in search of wisdom or truth. Debate is central to the Jewish construction of religious, political and social institutions. In many ways it is the foundation upon which Israel has built the notion that ideas should be freely advocated and freely defended.

Much like Judaism, the Chinese people have a long, rich history of debate and learning through the process of investigation. During the Neo-Confucian times, ultimate loyalty was owed to no individual but instead to the ultimate harmony of all things. As such, counselors and ministers were not confined or stifled by the respect they needed to show to the office and authority of the emperor. Respect could in fact be shown by being courageous and pointing out flaws in their rulers or seeking to reform a ruler whose disputes or rulings they did not agree with. Similar to way Jewish Rabbis referred to religious texts, Neo-Confucian era politicians also referred to classical texts to back and reinforce their beliefs.

From "Greater Learning" Da Xue 大学 comes the idea 修身齐家治国平天下, which translates to:

> *Wanting to light up the bright virtue of all in the world, the ancients first put their states in order. Those who wanted to put their states in order first regulated their families. Those who wanted to regulate their families first cultivated themselves. Those who wanted to cultivate themselves first rectified their heart and minds. Those who wanted to rectify their heart and minds first made their intentions sincere. Those who wanted to make their intentions sincere first reached understanding. Reaching understanding lies in the investigation of things.*

In other words, for the family, state, party, and country to maintain harmony – the ultimate goal – we must begin by exploring and investigating the world around us. Through this process we can reach truth and understanding.

In modern China, debate is most often associated with the formation of a debate team or joining of a debate program that can teach critical thinking, communication skills, research techniques and listening skills. Students are given the chance to expand their worldview about policy and decision-making. Perhaps no less important is ability of participation in debate programs can bolster a student's ability to be accepted to an international university. The reason for this is perhaps best described by Yale professor, Minh A. Luong:

> *"Extracurricular activities like forensics are playing an increasingly important role in the college admissions as well as the scholarship awarding processes. Why? Grade inflation is rampant in both public and private secondary schools and test preparation programs are distorting the reliability of national standardized tests like the SAT and ACT.*

*According to the Wall Street Journal, college admissions directors are relying less on grade point averages and standardized test scores, and are relying more on success in academically related extracurricular activities such as speech and debate..."*

A Public Radio International radio segment from 2014 offered listeners an in-depth analysis of some of the reasons that debate and forensics teams were slowly becoming popular in China. Among students competing for acceptance to top foreign universities from the USA to Europe, debate is one thing that set these achievers apart from their classmates. Zheng Bo, the international adjudicator of the inaugural Shanghai International Debate noted that debate sparks critical thinking. In an educational system where "teachers are given absolute authority and students just listen and recite and remember…when it comes to something without a standard answer … that's creating a lot of trouble, because they [students] are not familiar with this kind of practice." Unlike traditional classroom interactions in the Chinese education system, debate programs are able to prepare students to think on their feet. By using debate to develop these critical thinking skills, Chinese students are able to better respond in scenarios where there is not necessarily one correct answer.

One Chinese student described his experience with debate as "eye-opening." He reported that China's primary education "teaches you to love your country, to be patriotic—but through debate, we see that even though you do not praise your country, it does not necessarily mean you are not patriotic." Although the inaugural Shanghai International Debate kicked off with only 100 students in 2014, a number of prestigious US universities, including Harvard and Yale, have invested in sending educators, teachers, speakers, and recruiters to China in order to form new debate teams across the country.

China is readily accepting that the process of questioning, investigating and debating can be both popular and educational. In recent years debate programs have become very common in China and are offered through educational enrichment programs for students in local and international schools across the country. There are now well over 1000 students participating in national debate tournaments in China. Some of the debaters are even accepted to the prestigious summer workshops with the Harvard Debate Council and return to China to lead the rest of their team to successes in national and international competitions. By attending these programs abroad during their summer studies, Chinese students are gaining an advantage in the quest to study abroad at foreign universities. In addition to the full English-language immersion, Chinese participants can interact with other students their age who are also at the top of their class and aspire to attend the same elite universities.

Debate has extended beyond the academic world to become a well-loved Chinese national event. The hit TV show, Strangely Amazing Debate 奇葩说, was released in late 2014 and it documents the search for a master debater who is not only the most eloquent speaker but is also capable of expressing the most unique viewpoints through debate. The show has already accumulated billions of viewers around the Chinese-speaking world. Now in its fourth season, the show has gained great fan base with the millennial audience. Its debate topics range from fun to controversial. Collected from Chinese Quora, Baidu Forum, hot topics in the news, blog hits, etc., viewers can even vote for the most debate-worthy topics and engage in the debate experiences as viewers.

Having reviewed the Jewish and Chinese traditions of debate, we must now find a way to apply the lessons of these traditions to the educational experiences of Chinese students today. We seek to answer the central question asked by Chinese parents and educators: How can we keep our children respectful toward their teacher while still enabling them to ask more questions, find their own ways to solve problems and think critically within the framework of our Chinese education system? Below are the main ideas and take-aways that have can be practically and reasonably applied within the Chinese education system:

1) <u>Enrolling your student in a debate team or program</u>: Beyond the creativity and innovation that can be found by leveraging the debate format, participating in a debate program can provide a competitive-edge in the pursuit of higher education outside China. As noted above, colleges and universities around the world are placing more and more emphasis on critical thinking skills. Almost every educational institution has a written mission of preparing their students to be well-articulated, able to speak freely and intelligently, and able to think critically for themselves. Debate and forensics (competitive speech clubs) teams find themselves filled with those who practice the same styles of learning as the Jews historically practiced with their own religious documents.

2) <u>Create opportunities for "collaborative discovery"</u>: Debate within the Jewish tradition inherently requires a sparring partner, a friend. The combination of Chinese students often being the only child and having extremely busy schedules of academics and extra-curricular activities leaves little to no time for the Havruta experience found in Jewish forms of debate. Additionally, debating at home with parents can be viewed as argumentative and is not encouraged. As a result, Chinese students who are not part of a debate team miss out on the unique experience of "collaborative discovery" and creative discussion style. Chinese parents and educators can address this issue by creating environments both at home and at school in which a child can experience the possibilities inherent to group study and debate. Rather than being forced to constantly compete against one another, havruta debate style can enable Chinese students to create a partnership in learning. In the process, they are likely to discover new viewpoints and gain a deeper understanding of their ideas.

3) <u>Debate without rules or limitations</u>: Academic debate teams and competitions are often very methodical and executed with clear topics, parameters and time limits. While this can be helpful for students to develop critical thinking skills, it often lacks the spontaneity and impassioned exchange of ideas commonly found in traditional Jewish debates. In order to force Chinese students to truly think outside the box, explore opportunities for them to advocate for their point of view without any limitations, truly trying to convince their peers that they have used the investigative process to reach the truth.

4) <u>Practice Empathy</u> – A central part of the debate process is listening to those with whom you may disagree and gaining a true understanding of their perspective. By engaging with peers who see the world differently than they do, Chinese students can nurture their ability to empathize with others and see a problem from multiple perspectives. This skill will only become more important as we become more interconnected as a global society. Being able to understand different people's points of view is a critical part of succeeding as a global citizen.

5) <u>Use self-Investigation to gain clarity</u>: The traditional, ancient Chinese saying states that "clarity comes with debate". However, when we reflect on the practices of the formal education system in China we can see that this practice of self-investigation and the search for clarity has been obstructed. Chinese students must regain a mindset of debate and questioning within the respectful parameters of daily education. It is not impossible to continue to have a respectful learning atmosphere with the highest respect to the teacher and debate at the same time. Chinese educators are many times granted the same level of respect as parents and grandparents. Chinese students are courteous, do not interrupt easily, and don't challenge the authority. But students must take their success and future into

their own hands - within the framework of the current system. Students should be encouraged to engage their teachers from respectfully and make sure that no one is put on the 'defensive'. A student should always feel that they have the freedom to engage their teacher and their fellow students on the quest for truth and knowledge.

# Chapter 9: "Chutzpah" - Heedless Courage and Insight

It gives me great pleasure, indeed, to see the stubbornness of an incorrigible nonconformist warmly acclaimed.

-Albert Einstein

<u>Taking on Industry Giants</u>: a personal story from the author
*Ami: As an Israeli child, I was naturally full of "chutzpah" - the Hebrew and Yiddish word for audacity that's generally associated with courage and making gutsy moves. Throughout my life this part of my personality served me well, but in 2008 it brought my success to a whole new level. At the time the development of 3D technology was focused mainly on the movie industry. While everyone was working on the 3D cinema experience, through the company I had founded I discovered a new way to present 3D movies on TV. This was a completely new market concept, we were a very small company and I knew that in order to succeed, I would need to partner with one of the industry leaders. The TV market was dominated by huge international corporations like Samsung and LG, Panasonic, and other tech and media giants. And yet, I wasn't deterred. I knew that our solution was the best and I wasn't afraid of taking on the big guy. I knew I had to take a chance and make my voice heard. I quickly turned my company's solution into a product and built a prototype that I could exhibit at one of the major trade shows. The*

demonstration of our disruptive technology immediately drew the immediate attention of Panasonic. At the time, Panasonic had been moving the same direction trying to develop a consumer 3DTV technology that could be used with home entertainment systems. It was not long before we had established a close and fruitful partnership that leveraged their prowess as a media technology giant and our chutzpah as a small, innovative company.

Against all odds we committed to developing our solution and getting it ready for the global innovation stage, otherwise known as the annual Consumer Electronics Show (CES). In January of 2009, we arrived at CES to prove to the world that 3DTV was indeed a reality. The 3D film market was still in its early stages, with movies like Avatar scheduled to be released later that year. No other companies had shown real promise or interest in developing this technology for the home environment. After our CES demo the industry saw what was possible and it completely changed the landscape of the market. Within one year, EVERY single company, from well-established industry giants to production conglomerates - had all developed a 3D product offering. TVs that were rolled out over the next five years were all 3D ready. The TV industry experienced great success as consumers sought out these new products that brought with them the promise of the future of the home entertainment experience. What we could not predict, was the way this technology would become a passing fad. As we look around today, we can see that a relatively small percentage of the population is watching 3DTV at home. Despite its initial achievements, in hindsight you can say that our technology actually failed to revolutionize the industry the way we wanted. However, at the time, this joint venture with Panasonic proved to be both a business and financial success for me and my company.

More than our disruptive technology or our fortunate market timing, I attribute the early success of this venture to chutzpah. The entire journey began as an idea I had, a spark of creativity. I simply connected my computer and created a short 3D-TV demo. Once my idea had some traction with others in my professional circle, we started moving. This chutzpah gave me the confidence to believe that an idea that began in my living room at my old Tel Aviv apartment could really change the world. My chutzpah also fueled my next steps, enabling me to ignore the fact that I was a small fish in a big ocean and instead pursue a partnership with Panasonic. Here I was, the founder of a tiny startup, convincing one of the largest tech giants that I had the ability to completely disrupt and revolutionize the TV market. Sometimes I still find it hard to believe that my idea evolved from a small project to an industry standard for the TV giants of the world. Although there are a limited number of people watching 3DTV at home today, I view this venture as a major achievement. Most people may not know that our electric 3D glasses served as the cornerstone for what has become today's booming VR & AR industries. Our ability to sell millions of glasses, miniaturize the electronics and optimize 3D and Dual-View technologies are all part of ongoing efforts to make entertainment and education both more personal and more immersive.

I'm extremely proud of this chutzpah. I consider it one of the central forces behind my successful career as a serial entrepreneur. My chutzpah is the reason I am able to trust my gut, believe in my creativity, and do whatever it takes to make my projects take off. Without it, I would not have approached close to half of the people and companies I pitched my ideas to over the last two decades. Luckily, this chutzpah was cultivated in me from a young age. Rather than feeling like I needed to muster up the courage to take action, this confidence and audacity was simply part of who I am. It brought me to where I am today and I'm sure it'll play a critical role in all of my future ventures.

One of the most ancient stories of the *Midrashim* — the Jewish Rabbis intellectual interpretations of the Old Testament — is that of Abraham and the idols. This story is taught to young children and its message and themes convey yet another way in which the Jewish people value individualism and encourage people to question societal norms. The story begins with the ancient peoples who believed in Gods that represented the "heavenly lights," including the moon, stars, and sun. Idols made of clay and wood were created to embody these Gods. There was overwhelming consensus among the people of Babylon about the multitude of God-like figures they worshipped and served these idols faithfully. In the city of Ur in Babylonia lived a skilled idol-maker named

Terach. He made a living by selling idols of the gods at the marketplace, and often entrusted his store to his oldest son Abraham. Ever since he was a young boy, Abraham was considered rebellious. He was very inquisitive and proved to be an independent thinker. Rather than follow his father's tradition of idol worship, Abraham questioned who really created the heavens and the earth. He could see how the sun shone in the sky all day and at night was replaced by the light of the moon. He wondered how each could be so powerful and yet be replaced by the other every day and night. He began to question how these two separate entities could be considered Gods. Instead, he thought that people must worship one God to that had created all of these other forces. Abraham decided to worship only one God. As he shared his theories with those around him, he was greeted by a great deal of criticism and was considered rude and rebellious.

One day Abraham was entrusted to take care of his father's store in the hopes that he would eventually take over the family business. An experienced soldier entered and requested an idol who was a 'great soldier like himself.' Abraham gave him the fiercest looking idol in the shop. As the soldier was leaving he asked, "Are you sure this god is as fierce as I am?" Abraham asked the soldier his age. The soldier replied, "I have been a solider for more than thirty years". Abraham laughed and said, "My father only carved this idol last week, and now you seek protection from it!" The soldier stormed out of the shop leaving the idol behind. Next, a woman entered and said, "My house was robbed along with my idol. I need a new one." Abraham smiled and said, "Your god could not even protect himself, but you want to buy another?" The woman left the store.

Abraham's brothers reported to their father what happened. They suggested he be made a priest instead of a salesman. When Abraham asked, "What does a priest do?" the sons replied, "Priests stand before the gods, serving them, washing them, and feeding them." Abraham agreed to become a priest. He prepared food and drink for the gods and gave these offerings to the idols. As they were mere statues, they could of course not accept the gifts. Abraham famously described this scenario, saying of the idols: "They have mouths but cannot speak, eyes but cannot see; they have ears but cannot hear; noses but cannot smell; they have hands but cannot touch; feet but cannot walk; they can make no sound in their throats. Those who fashion them, who put all their trust in them, shall become like them." (Psalm 115: 5-8). Abraham took a stick and smashed every idol in the room except for the biggest one. He put the stick in the hands of the biggest idol. When Terach, Abraham's father arrived, he asked what happened to the gods? Abraham answered that they had all fought for the food offerings and the biggest one smashed the others and took the food and drink for himself. Terach, upset with his son, yelled that this could not possibly be true! And Abraham reasoned coolly, "Please let your ears hear what your mouth just said."

Abraham was considered rebellious by his family and by the society around him. In modern Hebrew, the word for his behavior is *chutzpah*. When it was first created and used, the word Chutzpah had a negative connotation. In Yiddish (the language of Jews throughout Europe before modern Hebrew was developed) it meant insolence or cheekiness. It usually describes a person who has gone out of the acceptable boundaries of action. But the root of the word from the Hebrew actually can be both positive and negative and refers to those who exhibit behaviors that are considered audacious, bold, and gutsy. Abraham went on to become what many scholars would consider the first Jew. Before him, the people of that fertile crescent of land were called "Hebrews." But God called upon Abraham to leave his home, challenging him to go out into the world and establish a new people- the Jews - who would only worship one God.

Abraham's chutzpah was his audacity to stand up to his family and the people around him. His chutzpah was the boldness with which he showed his father that his questions deserve to be investigated and answered. According to the Old Testament and the Rabbis interpretation of the texts, God rewarded Abraham for his spirit and chutzpah. As written in Genesis 12, God speaks to Abraham and commands him to: "Go from your country, your people and your father's household to the land I will show you. I will make you into a great nation, and I will bless you; I will make your name great, and you will be a blessing." As Jewish children are taught this story and its underlying lesson, they begin to internalize a central idea: being audacious, going against the grain, and essentially – having *chutzpah* – can reap great rewards.

In Dan Senor and Saul Singer's best-selling book "*Startup Nation*", the authors describe "*chutzpah*" as "gall, brazen nerve, effrontery, incredible guts – presumption plus arrogance". Senor and Singer continue: "An outsider would see chutzpah everywhere in Israel: in the way university students speak with their professors, employees challenge their bosses, sergeants question their generals, and clerks second-guess government ministers." To Israelis, the writers explained, this isn't something out of the ordinary. It's the normal mode of being. "Somewhere along the way – either at home, in school, or in the army – Israelis learn that assertiveness is the norm, reticence something that risks your being left behind." From the establishment of the Jewish people in the ancient kingdom of Babylonia to the flourishing innovation hub found in Israel today, the Jewish people made chutzpah part of their faith, education and culture. Jewish entrepreneurs with chutzpah know what they want and are willing to go for it. They will do whatever it takes to create something new and succeed. Chutzpah is the attitude that makes those that fail time and time again decide to get back up and try again.

Israeli Venture Capitalist and tech journalist Gil Kerbs encapsulates what it means to have Jewish chutzpah and an Israeli entrepreneurial spirit by describing the following scenario: "If an [Israeli] entrepreneur saw Warren Buffet, he'd immediately go to introduce himself, sure that his company is the exact thing Buffet is looking for, although Buffet never invests in technology or startups – that's chutzpah." Kerbs continues, "If [he] hears of a medical problem that nobody solves – he'd go at it, sure he can solve it – even if he has zero experience in the medical world and little to no understanding of biology – that's chutzpah. And when you think of it, chutzpah is a great thing to have as entrepreneur. When you are not afraid to tackle the hardest challenges and when you are shamelessly taking every measure to make it happen – you are more likely to achieve success."

In recent years Chinese culture and business leaders have taken greater notice of this aspect of Jewish and Israeli entrepreneurship and have sought to gain a greater understanding of what it means to have chutzpah. The following headline from 2009 is considered the first formally published Chinese translation of chutzpah: *wusuoguji de danshi* 无所顾忌的胆识. In Chinese chutzpah is defined as heedless courage and insight. While this definition was provided by a now defunct quarterly magazine from the Beijing based culture group, the Shao Foundation, the quest for understanding and integrating chutzpah into Chinese business practices is far from obsolete. More than ever, the Chinese economy is searching for ways around the hierarchical institutions of the past that value modesty and restraint over zeal and a 'startup' attitude.

Tencent, one of China's most well-known and most successful companies, can be used as an example of how chutzpah can lead to innovation and business success in a world of fast competition and mass media markets. Despite already leading the Chinese internet industry, in 2010 Tencent recognized the importance of developing a mobile-based arm of their business. Even though they weren't struggling to capture more of the market, Tencent decided to make the bold decision to compete against their own product. Leveraging the chutzpah of a small group of engineers inside the company, they began developing a messenger app that would rival their own existing platform, QQ, which had at that point amassed over 650 million monthly users. Over a period of three months, this team of seven engineers asked all of the right questions, used all information available to them, and had the audacity and gall to try and build something bigger and better than QQ. Other Tencent teams likely wondered what this small team could possibly accomplish? They had only with a small staff with limited resources, and yet, in three short months this team launched what would become the most influential China mobile app - the first version of WeChat.

Instead of settling for the user-friendly copy of existing mobile messaging apps that they had created, Tencent continued to question what was possible and probed further. They knew that to ensure the platform's mass adoption and longevity, they needed to integrate new technology that would make their product truly unique. This team of chutzpah-driven engineers found their answer in voice messaging. WeChat quickly evolved into an app that allowed users to send short voice messages to each other. WeChat continued to evolve and add unique features that went beyond voice and text messaging, including bill-pay, audio/video calls, and subscriptions to corporate accounts. This tradition of constantly evolving has continued until today, with WeChat consistently adding new features and capabilities, making it one of the most popular mobile apps in China. Tencent's success is in many ways a result of the chutzpah of these few engineers who were committed to taking on the impossible and innovating quickly and creatively. While they are not alone, there are a limited number of Chinese business leaders that have fully understood the value of chutzpah and have tried to integrate it into their business activities. The benefits of chutzpah may have particular importance in the lives of entrepreneurs, this audacity and willingness to make gutsy decisions can benefit individuals across all sectors and professions.

Even as students, chutzpah can enable individuals question the things they learn about in order to gain a deeper understanding of the subject matter. Likewise, chutzpah can be the reason a student stands out among a crowd of applicants for a foreign study program or international competition. As such, we must consider how China can cultivate chutzpah among Chinese students today, in hopes that they can leverage this boldness of spirit in order to change the world of tomorrow. How can students in the Chinese educational system, an environment that places so much value on respect for teachers, mentors, and elders, accept the expression of chutzpah without deeming it as a disrespectful? Below are some of the ways in which Chinese parents and educators can begin to teach the value of chutzpah and enable students to develop their own way of using chutzpah to achieve their goals. We are hopeful that by using such tools and practices, chutzpah can one day be considered just as commonplace in China as it is among Jews in their culture and traditions.

1) <u>Redefining chutzpah as tenacity</u> – As long as having chutzpah and acting boldly is viewed negatively within Chinese culture, it will be impossible for Chinese students to feel comfortable using chutzpah in their self-expression and interaction with others. By redefining chutzpah as a way of having tenacity and determination to succeed, it will be easier for Chinese parents and educators to accept the chutzpah of their students as something positive and it will also make these same students feel proud of their ability to act with chutzpah as they pursue their dreams. Providing students with examples of Chinese business leaders like the engineers of Tencent that were able to leverage their chutzpah will also contribute to the positive lens through which chutzpah is viewed.

2) <u>Opening up the room for questions</u> – Questioning the status quo is a central part of what it means to have chutzpah. By creating an open and inviting environment both at home and in the classroom, Chinese parents and educators can give students the courage to ask a question without feeling as though they are acting disrespectfully

towards their elders. Questions that come from a desire to discover new information are inherently respectful and are an expression of a student's inquisitive nature. This inner curiosity is precisely what we should be seeking to cultivate among Chinese students. Chutzpah allows one to delve deeper and is often the first step on a path of innovation. Just knowing that questions are encouraged can create space for chutzpah to develop.

3) <u>Finding the modern, Chinese iteration of chutzpah</u> – We can appreciate that the Israeli version of chutzpah may feel a bit brazen within the Chinese context, causing students to feel like 'know-it-alls', stubborn or rude. Finding a way for the core tenets of chutzpah to be integrated into the more moderate, polite interpersonal interactions among Chinese individuals will turn chutzpah into a tool for students rather than a risky behavior they might want to avoid. Perhaps this means identifying the most suitable situations for Chinese students to practice acting with chutzpah. Alternatively, parents and educators in China can model their version of chutzpah for children, showing them that they too are capable of acting audaciously when it can be used to serve a purpose. In this way, chutzpah can be integrated into Chinese culture without damaging the enormous respect people have for their elders, their teachers, their party and their country.

4) <u>Use chutzpah to find purpose and ignite passion</u> – As we examine the world around us it is easy to see that most leaders and global 'game-changers' have some chutzpah-like characteristics. These individuals have such a strong belief in what they have to offer that they no longer see it as risky or brazen to do whatever they deem necessary to succeed – even if that means making gutsy choices and acting boldly. Developing chutzpah can be life changing as it removes limitations and opens up new possibilities. It can inspire individuals to discover new talents or a sense of purpose within a particular profession. It can give individuals the courage to pursue their passions, regardless

of how these passions might contribute to their academic success or overall career goals. Rather than trying to create the perfect student by curating every course of study and extracurricular activity, Chinese parents and educators can encourage students to explore whatever they are most excited about with a bold and courageous attitude. In the process, students will learn how to use chutzpah to access their inner innovator. The benefits of integrating chutzpah into a student's life are both immediate and long-term. Students with chutzpah will seek more than academic excellence - they will view their education as part of a life-long process of developing their ideas and furthering the causes they care about most. Students can internalize chutzpah and use this audacity to elevate their success across a multitude of educational and business environments.

# Chapter 10:
# Risk Taking

Life is like riding a bicycle. To keep your balance, you must keep moving.

-Albert Einstein

<u>Taking a Leap into the Unknown:</u> a personal story from the author

Jordan: When I was a doctoral student in New York City, it would have been most comfortable for me to stay in my social circle with other Chinese students. Most students of particular ethnic, cultural or religious backgrounds chose to stick together. They studied together, dined together, and even lived together. With the multicultural, uniquely diverse backdrop of New York City, many chose the easier path of staying close to others who labeled themselves the same way: Chinese foreign students. Unlike many of my peers, I was determined to expand my social circle and step out of my comfort zone. I was driven by a deep desire to find out what I was really capable of. It was clear to me that when you don't know what you don't know, you settle for what you do know. I knew that I had to begin taking risks, both professionally and personally, and so I took the plunge.

During my second year of my doctoral studies, I applied for a position at the National Committee on US China Relations (NCUSCR). Despite not being sure I'd be a perfect fit, I was excited to be invited for an interview. Many years later, I still remember that day very vividly. The NCUSCR office, located in midtown Manhattan, had a red door and as I rang the bell I noticed a word written in German rather than Chinese. A gentleman opened the door for me and as I waited I began chatting with him. I asked why the doorbell was labeled in German instead of Chinese. He said he never noticed it, even after working there for six months. I soon went in for my interview and towards the end, they requested that I speak with the President of the organization. As I pushed the door to the President's office open, I saw the gentlemen who had greeted me at the door and whom I had previously asked about the doorbell. He wanted to let me know what he had found out about my question and proceeded to share that because the doorbell was made in Germany and the label hadn't been removed, it was still in German. While you wouldn't expect a Chinese candidate to ask very many questions in a job interview, this exchange about the doorbell was the start of a two-hour conversation where I inquired about his upbringing, education, work experience, how he got involved with the NCUSCR and his views on the Sino-US relationship. I later learned that rarely had he come across an interviewee who dominated the conversation with such enthusiastic and somewhat surprising questions.

To this day, I don't know where I gathered all my courage to ask so many questions to someone so obviously above my level of authority. I took a risk. I didn't follow the norms of my Chinese culture. I simply let my genuine curiosity shine through on my quest to find out how we end up doing what we do and become who we are. Although I didn't end up working directly for his committee and instead took a position at China Institute to run its Children's Chinese Program, I started to attend events organized by the Committee and leveraged these opportunities to meet people from all walks of life. I met colleagues that helped me come out of my shell, introducing me to an exciting world of leaders, risk takers and international role models.

My eagerness to step outside my comfort zone continued to pay off. At the time, I could have never dreamed that I would become friends with China expert and former Wall Street financier Ken Miller. My new attitude towards risk-taking led to my discussing education with former President of Yale University, Benno Schimit, and former President of Cornell University Jeffrey Lehman. It was not long before I found myself spending one of my birthdays at the residence of former mayor of NYC, Michael Bloomberg. No less exciting was the time I hung out with Sue and David Rockefeller Jr. during a fundraising event and learned about their charity work which included funding the first medical school in China more than a century ago.

These first leaps into the unknown would eventually lead to my learning what it takes to start and run a successful start-up with serial entrepreneur, Jeff Bogatin. The evolution and development of my risk-taking abilities led to my speaking with former CEO of IMG, Mike Dolan, about how degrees and majors may have little to do with what we can achieve at work and in life in general. It was the reason I had a chance to hear former Secretary of State Dr. Henry Kissinger (Nobel Prize Winner) talk about China. My curiosity led to my asking former Prime Minister of Australia Kevin Rudd how he learned Chinese, commenting in the process that his Chinese was easier to understand than his English. During my tenure at Teneo, I leveraged my risk-taking nature in order to pitch former CEO of Bloomberg LP., Daniel Doctoroff, on China's strategic plans.

While there are many to choose from, perhaps one of the most interesting experiences I had as a result of my willingness to take risks was winning a few ping-pong games against Hollywood movie star Alan Alda. Completely unaware of his status as a movie star, I laughed at his jokes as we played and will never forget how he accidentally pronounced his movie Four Seasons in Chinese as "dead chicken" when it was premiered in China. Letting my curious nature shine through, I picked Roger Lowenstein's brain on how he started a charter school in L.A. and how he raised a fearless and entrepreneurial son who now owns restaurants, bars and sports teams merely in his early 30s. I knew better than to be shy when interviewing renowned Harvard business professor John Quelch on his experience in running one of the best business schools in Shanghai and his view on innovation with Chinese characteristics. All of these amazing experiences began with my commitment to taking risks and leaping into the unknown. Building off of one experience of trying to work outside my wheelhouse, I spent the years to follow pursuing my passions and avoiding the traditional life of a Chinese student who only speaks when spoken to. I knew that I was better off making waves.

Risk-taking has always played a critical role in Jewish history. This is particularly true when we look at the numerous periods when Jews lived under siege. In Ancient times, Jews risked their lives fighting for their freedom against the Persians, Greeks and Romans. The same is true in more modern times when Jews fought against the Ottoman Empire and British Rule over the land of Palestine, prior to the establishment of the State of Israel in 1948. There are countless examples of the Jewish people facing the greatest risk - life or death, survival or annihilation. The response has become almost inevitable: innovate to survive and then innovate to succeed. Once you have risked your life, taking another leap into the unknown through scholarly innovation and creativity may not seem like that much risk at all. According to Steven L. Pease, author of *The Debate over Jewish Achievement*, nonconformity and risk-taking have been instrumental to the entrepreneurial success found within Jewish communities. Pease writes, "for 2,000 years, Jews have faced threats, and those helped shape their culture in important, almost Darwinian ways. Thus, the threat and adversity helped contribute to a positive outcome."

Rabbi Benjamin Blech describes this experience through his writings about one of Judaism's most famous figures, Rabbi Akiva. Blech writes:

*Long ago, the land of Israel was ruled by Romans, who enacted cruel and barbaric laws against the Jewish people. Once the government of Rome issued an edict forbidding Jews to study and practice the Torah. Despite this, Pappus, son of Judah, found Rabbi Akiva sitting in a public space, with students all around him, teaching and studying Torah in defiance of the Roman law. The penalty for breaking this Roman edict was death. Pappus was shocked that Rabbi Akiva was taking such a risk. In amazement he asked, "Akiva, aren't you afraid of the Roman government? For your life"?*

*Rabbi Akiva replied with a parable: Once, a fox was walking alongside a river. He could see fish swimming in schools in the water. It appeared to him that they were swimming to and fro, as if trying to escape something or someone. The fox was very hungry and thought that a nice, fat fish would surely make the delicious lunch for a hungry fox. The fox called out to the fish, "What are you fleeing from?" the fish replied, "We are trying to avoid the net that the fishermen cast to catch us."*

*Slyly, the fox said, "Would you like to come up onto the dry land so that you will be safe from the fishermen's nets?" The fish weren't fooled by the fox. They replied - "if we are in danger here in the water, which is our home, how much more would we be in danger on land where we cannot breathe!"*

*Rabbi Akiva continued, "We fear our enemies and take risk here when we study the Torah, but if we abandon our study, which gives us hope and life, we will be more fearful still." The two men continued to walk together through the villages and towns and together they risked their lives in the name of the education of the Jewish people. They brought the study of Torah to the forefront and led the way despite the greatest risk - that of death - being imminent.*

This parable can be translated and applied to many aspects of our lives. What is most important is to remember that taking a risk - even though it can require a leap of faith - can also yield the greatest rewards for those who are passionate enough to see their vision through. In Jewish Rabbinical teachings and interpretations, it is often said that "Every descent is for the sake of a future ascent". More simply put, every negative (failure) will eventually lead to something positive (success).

Another example of this instinct to take risks and innovate as a means to survive can be found in the life story of Nobel Prize Winner, Roald Hoffmann. Hoffman and his family were living in what is now Ukraine when World War II began, and they were forced into Nazi labor camps when Hoffmann was just a child. Miraculously, Hoffman's family was able to bribe guards to allow his mother, uncles and five-year-old Roald to escape what was certain death. Having to leave his father behind, Hoffmann and his mother escaped to the attic and storeroom of a local school where they remained for a period of 18 months. While in hiding, they faced the daily threat of being discovered. Hoffman later shared how his mother kept him entertained by teaching him to read, making him memorize geography from old textbooks and then quizzing him on the knowledge he had acquired. Hoffmann referred to the experience as "being enveloped in a cocoon of love". He goes on to write, "you would think that the simple act of quizzing me on geography would be taken as a boring daily routine. But when you can't go outside for fear of death - the learning becomes a different experience." In addition to many other national and international prizes of merit, Hoffmann was awarded the Nobel Prize in Chemistry in 1981, sharing the honor with Kenichi Fukui "for their theories, developed independently, concerning the course of chemical reactions." Beyond his work in the field of Chemistry, Hoffmann is also a published playwright and poet. In an interview with *Scientific American* he defined his research style as distinctive: "I don't start with big tasks or the great questions of chemistry, I do many small problems inspired by experimental work…Everything in the world is connected to everything else. I know I will begin to see connections." It is this type of thinking that a risk-taker, one who has faced the threat of death, can use to create truly innovative solutions.

Whether it is Rabbi Akiva protecting the right to study the Torah during the Roman Empire, or Nobel Prize Winner Hoffmann studying geography in a secret attic to avoid capture by German Nazis, taking risks for both personal education and the greater community's enrichment has always played a prominent role in Jewish history and culture. Anything will be done in the name of education. Failure, even in the face of death - isn't acceptable.

Before we delve into the lessons that can be learned from the role risk-taking plays in innovation among the Jewish people, we must first gain a better understanding of the obstacles and challenges that Chinese students face on the path to becoming more risk-takers and non-conformers. The first element of traditional Chinese culture that can influence an individual's willingness to leap into the unknown is the immense weight of responsibility they feel to succeed. Since the 1970s and China's one child policy, a Chinese child was born and raised with love and cared for by as many as six adults. From the day they are born, Chinese children, their education and their future success are of the highest priority for Chinese parents and grandparents. Families will invest endless time, effort and resources into providing their child with the best possible education and related learning opportunities. Parents and grandparents consider child-rearing almost a 'second job' and will devote whatever is needed in order to teach and prepare Chinese students for schooling, and examinations. Children might feel loved, but they would also feel a great deal of expectation and responsibility, understanding that the pride of the whole family sits squarely on their shoulders.

With this excessive emphasis on academic performance and little room for discovery, imagination or play, Chinese students can often become robotic study machines. They spend the majority of their lives making very few choices on their own, feeling stifled by what they 'must' do and lack the agency to control their own futures. Many students complete their entire education without ever taking any risks or stepping outside their comfort zone. Rather than pursuing opportunities about which they feel passionate, Chinese students often remain committed the paths chosen for them by their parents and fulfilling the wishes of their families. As a result, many Chinese young adults feel lost when entering adulthood, even after graduating from exceptional colleges and universities both in China and abroad.

The second and perhaps most important obstacle on the path to becoming a risk-taker is the fear of failure. We have mentioned this issue in previous chapters and its prominence in our review and analysis is perhaps reflective of the role fear of failure plays in Chinese culture as a whole. One of the most current and relevant examples is the Chinese start-up landscape. In contrast with the fast-paced, disruptive innovations hubs found in both Silicon Valley and Israel, Chinese high-tech companies are relatively very risk-averse and tend to play it safe. As a result, each industry is led by a few bold innovators and risk-takers that successfully introduce new technologies into the market, followed by numerous "copycats" that are able to leverage China's massive market in order to make a profit based on a more innovative company's proven success. These followers have not innovated, but rather provided a similar or slightly improved product design, or idea. As one can imagine, when the majority of Chinese entrepreneurs are looking to follow and copy rather than discover and create, China's potential as a leader in innovation is limited.

Fear of failure is equally prevalent in China's education system. While the focus on exams and competitive scoring exists was originally designed to give equal opportunities for all Chinese students to succeed, in reality, not everyone can or will achieve the grades needed to land them a spot in college or university and for many, failure to reach the incredibly high standards set by their families is the most likely outcome. In 2017, Harvard accepted only 5% of its more than 30,000 applicants. The math is simple: with over 24 million Chinese high school students graduating every year, all vying for limited opportunities to continue their higher education, there are bound to be millions of Chinese graduates who enter adulthood fearing the very leaps of faith they must make in order to find a new, alternative path to success.

From our brief review of historical risk-takers, we can see how the ability to explore the unknown has often yielded the utmost and highest levels of success. The 'great' ones are remembered for the times they went against the grain and those that merely followed are often forgotten. The question then for Chinese parents and educators is how to nurture risk-taking while ensuring that Chinese students succeed within the existing education system. Below we offer some of the most important lessons and how to apply them in order to encourage and promote risk-taking among Chinese students:

1) <u>Find joy in the risk-taking process</u>: In order to be truly committed to becoming a risk-taker, students must be able to experience joy along the way. If they are able to feel safe asking questions without judgement, they are much more likely to see their curiosity as a blessing and not a curse. Much like the examples provided in this chapter, by seeing the excitement and positive experiences that can result from taking risks, they may begin to associate becoming the 'Outlier' (as coined by Malcolm Gladwell) as a path to reaping many rewards. However, and perhaps more importantly, if students in China can enjoy the openness of new possibilities that is integral to the innovation process, whether or not they succeed, they will be more likely to continue risk-taking, regardless of the end result.

2) <u>Accepting failure as part of the process</u> – From sports to arts, science to humanities, there is no subject in the world that can be mastered without determination, practice, the willingness to take risks – and failure. Students can learn a great deal from seeing those they admire be willing to fall down, miss their target, and make mistakes. Invest time in demonstrating how these failures that appear so scary are merely opportunities to learn and grow as one navigates the road to success. Every single time an innovator fails on the way to a higher goal with a higher risk, they are 'failing forward'. By broadening one's focus, each 'descent' can be seen as a necessary part of future growth. When student, teacher and parent alike can approach failure with this perspective, they can together focus on moving forward towards success, leaving behind the "bump in the road" that did not go perhaps as well as planned.

3) <u>Stop thinking and take action</u>: Perhaps your student has decided that he or she is willing to take risks but is unable to take the next steps.

The most important way parents and educators can encourage such students is by demanding that they take ONE step forward. Many times, it is this first step that they fear the most, but the sooner students can harness the intrigue of innovation and the thrill of the unknown, the sooner they can discover their innovative spirit. Delaying allows more time for fear to grow. The most successful and innovative people are always ready to take a risk and often feel like they don't have time to wait. Cultivate this same sense of urgency and willingness to take risks by demanding action.

4) <u>Create a daily routine of positive thinking</u>: Scientists across the globe have spent years studying the effects of positive thinking. Here are two practices that can be integrated into your student's daily routine and that are designed to enable young minds to envision the future they see for themselves and take the risks necessary to realize their dreams. First, ask your student to complete daily observations with the following questions in mind: Have I taken a risk or done something new today? Have I taken a small step towards a bigger risk I'd like to take in the future? Are there any connections I can make between my actions yesterday and those of today? Second, have your student end each day by visualizing their ideal tomorrow. By leveraging the power of their unconscious mind during sleep cycles, students can help positive thinking become part of the fabric of their consciousness. Additionally, the ability to visualize success can often decrease the level of stress and diminish the fear of failure that students encounter along the way.

5) <u>Go the extra mile, even when nobody's looking</u>: In order to cultivate and develop the minds of future innovators, the motivation to succeed must come from the desire to change the world rather than the fear of academic failure. Nobel Prize winners weren't looking for a perfect score on a test; they were looking to solve problems of the world. Standing out in the crowd can stem from proving that you are

truly motivated and are willing to take risks to achieve your goals. Students and parents alike must be willing to alter their mindsets and replace fear with more positive motivators that reward curiosity and risk-taking. When students begin "going the extra mile" at an early age, it will become an innate part of their learning process. They will expect more of themselves in the day to day and not just when there are high-stakes, like on applications to prestigious high schools and universities. In doing so, the exams and rankings that loom so large and heavily influence the lives of young students are put into perspective. They are simply another obstacle to overcome on the path to something greater – innovation, creativity and success.

# Chapter 11: Military Service

I speak to everyone in the same way, whether he is the garbage man or the president of the university.

- *Albert Einstein*

<u>*Discovering My Inner Strength:*</u> *a personal story from the author*
*Ami: I was in my early 20's when, seemingly overnight, I became the Head of Security to the Israeli President as part of Israel's Secret Service. While at the time it felt as if I was thrust into this position completely unprepared, in reality I had been training for this level of responsibility for the previous four years as part of the Israeli Defense Forces (IDF). At 18 years old, my service began when I commenced the first level of the two-year training course to become a pilot. This is considered one of the most prestigious, grueling and intensive training programs within the IDF and I was extremely excited to be chosen among thousands to participate. The program was notoriously known for narrowing down its cadre of pilots to only an elite few that would make the cut and end up serving as pilots. Once I began training, I could completely understand why. As a pilot you are completely alone up in the air and are responsible for an aircraft worth millions of dollars. Yes, you receive training on the ground and yes, you are guided on your helmet radio by the voices down below. But at the end of the day, every split-second decision and every*

slight touch of the yoke would move the aircraft thousands of meters in the air, pitching and rolling in every direction.

Unfortunately, my excitement didn't last long. After several months of training, I was told that I hadn't made the cut and was not selected to continue on to the next level of the course. In that moment my disappointment took over me and it was the first time in my life where I felt that I had truly failed. But as the moment passed, I knew that I was destined for other things and began looking for the next mountain to conquer. Before long I was reassigned to a commander role in a tank unit. I became fascinated by tanks, both the engineering of them and the way they functioned. I started out leading a team of four in charge of a 72-ton tank. As I learned more about my leadership role and the way we functioned as a team within the tank, it struck me that this metal beast was in many ways similar to the human body. Our team needed to work as one, completely in sync, in order for the tank to function properly. In this way, I was responsible for the lives of my fellow soldiers, and they were responsible for mine. We all understood that any weak link in our group might result in possible death and so we learned to function like a well-oiled machine in order to protect both ourselves and each other. You don't see where you're going, and everything is unexpected. As I mastered this role I realized something important – a lesson that would shape my entire future – I was capable of leadership and this role is where I was meant to be. It felt right. Where I had failed as a solo actor, in the role of a pilot, I had excelled as the commander of a unit, capable of leading my team with absolute precision and complete confidence.

As a self-educated, child programmer that developed video games as a 12-year-old, my post-army dream was to study computer science and become an innovator in this space. However, reality had a different plan for me. Just when I was about to leave my military position behind and start my new job as an air marshal in order to make the money I needed for college tuition, my superiors asked if I would be willing to be reassigned. They wanted me to be part of the Security Unit for Yitzhak Rabin, Israel's Prime Minister at the time. My first day in my new role started only hours after one of the most tragic and consequential moments in Israeli history – the assassination of Prime Minister Yitzhak Rabin. The Secret Service needed to be rebuilt and all the officers *were to be re-assigned after what had happened. As Israel prepared for one of the largest memorial services in its history, I was tasked with securing international Heads of State from numerous nations that had flown into Israel for the funeral. From commanding a team of four, I found myself leading massive teams that could sometimes include thousands of people. Naturally, I was hesitant at first. "Do I have what it takes?" I thought. Despite the confidence I had gained, I was still only 23 years old. "Will I be able to rely on myself and on my team?" I wondered. I was responsible for ensuring sure that every one of my agents on the ground knew what their role was and could execute their jobs flawlessly. I didn't have time to reflect on what had happened and the complexity of this enormous security operation. I simply had to continue moving forward and focus on the tasks at hand. I had to rely on all of the skills that I had honed up until this point.*

*Despite the challenge, I innately knew that I could succeed. Unlike Prometheus, the weight of responsibility on your shoulders is not seen as something negative, or something that can crush you. Instead you focus on rising to the challenge. You simply act, knowing that the heaviness will dissipate as you move forward towards success. You perform as if you were born to do this and feel completely ready, even if you never thought you could. That's what my experience gave me. Before I had even chosen what subject I would study or what type of profession suited me best, I had figured out who I was inside and what it would take to succeed. Working with Prime Minister Shimon Peres as a young security agent and working as the head of security for Prime Ministers Netanyahu and Barak, as well as US President Bill Clinton, forever changed the way I view the world and made me see life with a different perspective. All of a sudden becoming the man on the ground than needs to make critical decisions in fractions of seconds that can impact global history, I learned to evaluate both my leadership skills and the leadership skills of others in a new way. I gained a deeper understanding of the meaning of risk, motivation, and responsibility. More than anything, I learned to question the realities around me again and again and to never stop seeking and searching for new innovative answers to these questions.*

Among Jews in Israel, there are two main rites of passage to adulthood – one religious, one cultural. The first rite of passage occurs when a young man or woman reaches young adulthood and celebrates what is called a "Bar Mitzvah" for boys and a "Bat Mitzvah" for girls. It is at this age that the Jewish religion considers these individuals old enough to take on the traditions and responsibilities of Judaism as adults. The second, more cultural rite of passage takes place at age 18 when young Israelis graduate high school and join the IDF or long-term, full-time volunteer programs in order to complete their mandatory military or national service. While a small percentage of high school graduates choose to participate in National Service volunteering programs, the vast majority of Israelis find themselves putting on a soldier's uniform and heading to basic training (boot-camp) within a few months of graduation. The following two to three years of their lives will perhaps be the most influential in determining who they are and what their future holds. Even before applying to university or holding their first full-time job, young Israelis are tasked with the greatest responsibility of their lives. Only a short time after their biggest concern might have been which beach to go to on the weekend, they find themselves making critical military decisions and taking actions that may save their lives or the lives of others.

While Israelis begin their military service at age 18, in many ways, they spend their entire childhood learning the skills they will need in the IDF. Beginning as early five years of age, Israeli kids are given a freedom and independence that is not common in other parts of the world. School buses were a fairly recent addition to the transportation system in Israel and many Israelis can recall taking a regular city bus, alone, in order to get to school. On the first day of the academic year, parents might tell their children: "you can meet your friends at the back of the bus, they know where to get off and how to get to school." This type of trust and independent attitude serves to boost the confidence of these young children as they navigate the world. They understand that while they are supported, they are also able to go out on their own and make their own choices.

This informal preparation continues throughout childhood. In middle school students participate in a variety of field trips to different areas of the country, accompanied by only a few adults. These trips include social, intellectual and physical challenges and can take place in the middle of a busy city or a desert crater. Many children are also part of the Israeli Scouts movement called "Tsofim." Children in the scouts program gather for weekly, student-run meetings and also attend week-long and summer-long camps where they are constantly taught to be more self-sufficient, self-motivated, and self-aware. Teens in Israel are quick to understand that they alone are responsible for their actions and responsible for their studies. They are given a chance to prove themselves constantly to the admiration of their peer groups. They are given support from the adult groups around them - parents, teachers, counselors - but they are never coddled and treated as being too young to make their own decisions or follow their own passions. The combination of these experiences enables young Israelis to prepare, year after year, for their future military service.

At age 16, Israelis will begin to be screened and tested by the IDF to begin identifying the most suitable military assignment for each individual. Many even choose to participate in rigorous physical training outside of their classroom just to be prepared for a more active duty placement in the IDF. When they finally graduate high school and completing all the exams and screenings for placement, they are ready to be considered soldiers. No longer seen as kids or students, parents, families and society as a whole now views these young adults as soldiers. Serving as a great societal equalizer, an IDF unit can be comprised of individuals from very diverse backgrounds, socioeconomic levels and academic abilities. A unit filled with wealthy sons and daughters of doctors and lawyers might be led by a unit commander of limited economic means who happens to be the son of an immigrant shoe repairman. Authority and rank is respected, regardless of who you are and where you came from. This also enables students with talents not tested in an academic setting to excel in a different way. In the IDF they can leverage their character, grit and smarts and walk away from the battlefield with a greater understanding of their abilities outside the classroom.

While the majority of men serve for approximately three years and the majority of women for approximately two years, soldiers who seek additional training and are recommended for higher-ranking positions will go likely serve for a period of four to five years. Those who complete Officer's Training school are given the option of serving as career military professionals and entitled to early retirement at the age of 40. For most this is just in time for them to pursue a second career where they can put their skills to use in the private sector. For those discharged from the military following their mandatory service, they must continue serving as reservists. This means that until age 40, depending on their role in the military, they will spend a few weeks every year completing training exercises. In addition, they can be called upon at any time if required in order to enable Israel to respond to national emergencies such as war or natural disasters.

While Israelis may have moments of wishing that they, like so many other young people in the world, could start their bachelor's degree studies sooner, the IDF is a school in and of itself. One of the central skills that military service in Israel is able to cultivate among young adults is quick-thinking. For many countries around the world, having an army mentality usually refers to a strict chain of command and the need to follow orders. While this is also true in the IDF, it is only to the extent that order is required for effective military execution. Unlike other militaries, in Israel even young soldiers and new recruits are very quickly tasked with delegating, making decisions with great impact and being responsible for yourself and others in your unit. As a result, Israeli soldiers are able to think on their feet, respond quickly to a dynamic situation and take responsibility for the outcome.

Another important life lesson Israelis learn in the military is a deep appreciation for true friendship and solidarity. Young soldiers are committed to one another and their united efforts throughout their service and for decades to follow. It is not a surprise that for many adults in Israel, some of their closest friendships are those made during their military service. Likewise, Israelis move through the stages of their life, their fellow soldiers serve as an important network for them to pursue their dreams and accomplish their goals. The people a soldier serves with at age 18 often turns into their future travel buddies, roommates, business partners and fellow innovators.

As noted above, the IDF also provides individuals with an important non-academic or classroom-related opportunity to explore their strengths and values. More than any aptitude test or academic score, the IDF can show a solider his or her true capabilities and interests. Young adults learn to push their own limits, no longer accepting their previous conceptions of themselves as the truth. Some prove that they can run 50km wearing all of their gear in the middle of the desert in August. Others learn that even when it appears that there is no solution, they are capable of doing whatever is needed when a life is on the line. It is in these real-life scenarios, often with great consequences, that Israelis are able to discover their inner strengths and perhaps reevaluate what they thought they were capable of. As they leave the military and move on with their lives, these individuals can then apply the strengths they uncovered towards achieving whatever future goals they may have.

While young soldiers are given a great deal of responsibility, they are also not expected to do more than their best. It is expected that they will not always choose correctly and that they will need the guidance of their commanding officers as well as the diligence of ongoing practice in order to improve. The preference the IDF has for letting soldiers fail on the way to success is more than just a particularly progressive approach to military training. It dates back to the Biblical figure of King Solomon, known in Judaism as Solomon the Wise, who wrote the following about failure: "The wise man falls seven times and rises again" (Proverbs 24:16). This ancient Jewish wisdom rings true even for young adults in Israel today and it places the focus on how an individual chooses to respond to failure rather than the fact that they failed in the first place. By encouraging soldiers to learn from their mistakes rather than fear them, they are able to overcome a great many obstacles. Given the critical role failure plays on the path to innovation, Israelis are more prepared than most for the failures they may face as innovators. They have spent their years in the IDF learning from their mistakes and – both physically and mentally – getting up, dusting themselves off and trying again. In understanding the role the military places in cultivating Israel's hub of innovation, we now ask: How can we integrate this uniquely "Israeli" experience into the lives of young Chinese students? Chinese classrooms and for young Chinese students?

1) <u>Prioritizing Team Sports & Activities</u> – Camaraderie and the feeling of working as a team are central to the Israeli military experience. While there is a clear benefit of excelling at individual activities like tennis or chess, parents and educators in China should also incorporate group activities that require young students to learn the value of working as a team. For example, encouraging students to play a team sport like soccer where they can cultivate the social skills of navigating physical and mental challenges in a group setting. Each member of a team is responsible for himself, however the team only wins if they can play and work together as a united front.

2) <u>Build Confidence</u> – One of the main take-aways Israelis have from serving in the army as young adults is that it instills confidence. Confidence affects not only a student's performance but the experience they have as they perform. With confidence, a loss may

seem like just another bump in the road to better skills, better performance, and the eventual win. The best professional athletes know that talent isn't everything and that in order to win they must have an internal sense of confidence in their abilities. Building confidence, like any skill, is a long-term process. By neutralizing fear of failure and focusing on the small, but important achievements students reach along the way, parents and educators can help cultivate a student's confidence. Establishing an ongoing custom of encouragement to try again rather than discipline for failure can change a student's self-perception. Doing so enables them to leverage their confidence in order to visualize, and eventually realize their success.

3) <u>Learning from failure</u> – In our chapter on risk-taking, we touched on the importance of learning to accept failure as part of the process. However, what we can learn from the culture and traditions of Israeli military service is the importance of going beyond acceptance and actually learning from one's failures. Rather than trying to ignore the failure or only focus on disciplining students for failing, parents and educators can strive to help their student learn from the experience. Try studying the failure the same way you would a scientific experiment: what were the causes? Rather than view the scenario through the eyes of shame and defeat, students should aim to see the experience from a non-judgmental perspective, and ask themselves honestly: what can be changed to avoid failure the next time? Anything from extra tutoring to a change in the daily schedule could facilitate future success. The more open parents and educators are to learn from their student's failures, the easier it will be to work together to find the path to success.

# Chapter 12:
# Travel, Learning & Destiny

"I have no special talents. I am only passionately curious."

*-Albert Einstein*

<u>Arriving into an Adventure:</u> a personal story from the author
Jordan: *My very first trip to the US was not short on drama. As a foreign student, the single most important legal document that gives an individual permission to enter the US is the I-20. This form was obtained through much paperwork and proved to the US customs and border patrol that I was legally enrolled in an academic program at Columbia University in the United States. Without it, I would not be allowed into the USA. As you can imagine I was thrilled when I received it from my future alma mater – I was ready! When it came time to begin my journey, I said goodbye to my family and friends, headed to the airport and boarded my flight. I was scheduled to have a short layover in Chicago on the way to New York City where I would begin my studies. Needless to say, I was full of many emotions – excitement, fear and a bit of courage for good measure.*

*Somewhere between getting off of the plane and waiting in line at customs, I had the awful realization - I had misplaced my I-20! Here I was, all alone in a strange country and I had lost the only proof that it was legal for me to enter the United States. To make matters worse, I had landed in the Chicago only a* short time after the 2003 outbreak of SARS in China and the US customs office was unlikely to make any exceptions to their very strict travel regulations. As I searched my belongings, I realized that in addition to my missing I-20 document, I hadn't brought any record of my plans for living and studying in the US. After relaying all this to the law enforcement officer, I was sent to a dark room to wait. And I did just that. One hour passed, then two, then three. After taking three full hours to complete my background check I was finally approached and told to speak to one of the other officers. I was exhausted and disoriented after my 13-hour flight and dizzying time at the airport, but I somehow gathered my strength and explained the scenario. I asked the officer to please let me go and assured him that Columbia University could reissue the document for me as soon as I arrived in New York.

I had never left China or set my feet in a foreign country before, but at that moment I had never wanted anything more than to be allowed to enter this foreign land and continue my journey of discovery. As I pleaded my case, I became deeply aware of what I might lose were this officer to decide to simply follow the rules and send me back to China. It was a long negotiation process, but in the end the officer finally agreed to give me two weeks to get the I-20 reissued and appear at the main offices in Washington DC to present my documentation. I was extremely relieved and proud of myself, but this sense of relief didn't last long. By the time I exited customs, I had long missed my connecting flight to New York. It was past midnight and I was stranded in Chicago's O'Hare airport with my two huge bags. I had no phone with which to call my sister and let her know what had happened or reach out to anyone for help. I had only a few hundred dollars to my name after leaving my hard-earned teacher's salary to my parents. I promised them I would make a decent living in America. And yet there I was, feeling completely lost.

This feeling of helplessness soon turned to motivation – I would figure this out. I wandered around the airport trying to find someone whom I could ask for help and finally spotted a man in a uniform. I approached him and began to explain my situation. It turns out that he was also once a foreign student in a foreign country. Thankfully, he understood my situation and was willing to help me find a solution. Not only did he reach my sister by phone so that I could tell her what happened, he also helped me contact the nearby airport *hotel for passengers who missed their flights. Soon enough I had arranged a seat on a flight to New York in the morning, a room at the airport hotel to rest until the flight, and a shuttle bus to bring me to and from the hotel. I thanked the man in uniform for his assistance and went on my way.*

*The following morning as I landed in New York City and made my way to Columbia University, I realized that I had successfully weathered my first "storm" in the US! From this unexpected adventure that could have ended with me on a return flight to China, I learned that I was able to think on my feet and find a way to overcome the obstacles I encountered. Of course, misplacing the document had been my mistake – but everyone makes mistakes. The real test is what you do next. With determination, some courage and being willing to ask for help, I had found a way to stay in the US and get myself where I needed to go. I spent the following decade living and studying in New York City, travelling all over the US and the world for business and pleasure – all while maintaining a sense of ease and calm.*

*I believe that my experiences of international travel and study abroad have given me the kind of confidence that it takes to survive anywhere in the world. I know that as long as I am living and breathing, I'll be able to carve out my space in the world. Life is a journey and travel has made it possible for me to expand my horizons and live my life to the fullest. These international experiences are also the reason why I devote so much of my life to educating and guiding Chinese students on their study abroad planning and life-long growth. Through mindful travel, study and learning, we are able to realize our strengths and discover our passions. To discover the world around us is to create opportunities for fulfilling our own destiny.*

There has been much research done on the subject of travel and innovation. The results of such studies have shown time and again: travel increases opportunities for connecting with new people with differing ideas and beliefs, and as a result, individuals who travel are more willing to be open-minded and are able to be more creative. Travel is nothing less than an investment in yourself and your ability to be innovative. It therefore will likely not come as a surprise that in the analysis of a cross-section of Jewish Nobel Laureates, we revealed an important commonality among them: they all left home to pursue an international experience. For those who left by choice, they most often chose to complete their high school or university studies at centers specializing in their field of study. Some traveled during their PhD studies or lectured as visiting professors. For those who were forced to leave their homes, most often fleeing as refugees, they found themselves as new immigrants in foreign lands, forced to explore and learn from their new surroundings. Regardless of the circumstances under which they left home, these Jewish Nobel Laureates were intent on finding and joining research institutes that would enable them to take part in the sharing and development of knowledge, new ideas, and innovative approaches.

In addition to the international travels of mostly European Jewish Nobel Laureates, there were many thousands of Jews who emigrated from around the world to the newly independent land of Israel. This included those who sought refuge from the war, those who saw Israel as the only place that would grant them religious freedom and those seeking adventure in the efforts to build up the country's infrastructure. Jewish immigrants to Israel were from a vast array of cities around the world, turning the young nation into a multicultural center rich with an abundance of languages, customs, and cultures all vying to find their place within the fabric of this new society. For the country to be able to grow, it required that the multicultural nature of its people be open to one another and cooperate. The result was a melting pot of new ideas that would serve as a catalyst for innovative collaboration and the birth of the hi-tech industry known today as "Startup Nation."

The tradition of Jewish international travel has remained strong within Israeli culture. In fact, one of the many rites of passage in the lives of young Israelis after completing their military service is to spend several months abroad. This 'post-army-trip' is both accepted and encouraged within Israeli society. Many of these young Israelis are transitioning from adolescence to true adulthood- and they know that their entire careers lie ahead of them. After completing two or three years of military service and prior to starting their undergraduate studies, young adults in Israel allocate anywhere from one month to a year to travel the world. According to 2013 Forbes article reporting on research conducted by Issta, Israel's largest travel agency, 30,000-40,000 Israelis backpack overseas every year. Seventy percent of these backpackers are between the ages of 20 and 24, which means they represent one third of the 75,000 Israelis who are discharged from the Israel Defense Force each year.

Of course, to those outside of Israel, this might seem like an excuse for just a long vacation, and perhaps a nice escape after a grueling army experience. But in reality, this time abroad is about much more than that and it is part of Israel's cultural infrastructure for innovation. During these trips, Israelis are able to completely relax their mind and body, opening themselves up to new experiences. They are studying at "the university of life," outside the classroom, in a way that fosters communication, creativity and independence. From arranging their own international flights and visas to finding themselves on a whim to visit a lesser-known village in whatever country they are visiting, these young Israelis are completely on their own. As described by the previously mentioned Forbes article, "60 percent of Israeli backpackers fly to Asia (mostly South and Southeast Asia), 30 percent to South and Central America, and the rest to Australia, New Zealand, and Africa. On average, they travel for six months, while ten percent travel for more than a year." Israelis are likely to find themselves anywhere else in the world, unable to speak the language, and with no set path ahead of them. They seek out every kind of adventure these exotic lands have to offer, including dangerous hikes the tallest peaks - testing their stamina, quick thinking, and physical skills. Such journeys are often followed by excursions to the sleepy beach towns and quiet enclaves where they can relax and prepare themselves mentally and emotionally for the next dimension of their lives.

Risk, creativity and innovation are at the heart of these travels. Within constantly changing environments, Israeli travelers must use their ingenuity in order to connect both with locals and other travelers. From food to accommodations, transportation to companionship – Israeli backpackers understand and accept that they must take risks and be willing to explore the unknown in order to truly enjoy the world around them. Furthermore, since the vast majority are traveling with the little money they were able to earn at odd jobs or a small bonus they received for completing the army, they must be thrifty and resourceful in order to maximize their travel experience. This includes finding ways to market their skills, work for short periods of time and barter whatever they are carrying on their backs in exchange for housing, food and transportation.

These transformative international experiences are often what enables Israelis to find their true selves – their strengths, weaknesses, likes, dislikes – and everything in between. Before even choosing the academic focus of their upcoming university studies, Israelis have the opportunity to see what the world has to offer and get in touch with their own creativity in the process. Upon returning to Israel to begin their studies, they have an easier time tapping into this creativity for inspiration and innovation. Travel provides these Israelis with a broader perspective and an array of experiences that can be applied to future fields of studies and, ultimately, their chosen path to success.

While this may be the most common age at which Israelis choose to explore the world, the passion for travel doesn't stop after this post-army-trip. Israelis have developed numerous technology platforms enabling them to leverage last-minute deals to fly abroad and it is becoming more and more common for Israelis, particularly in the hi-tech industry, to travel outside of Israel for business purposes. Following the 2013 Open Skies agreement ratified in Israel aimed at boosting airline traffic to and from Europe, international travel has also become much more accessible for all socioeconomic classes in Israel. Perhaps it is most telling to look at a recent statistic that showed that the Jewish population of Israel is approximately 6.4 million people and in 2016 alone, outbound air travel by Israelis reached 6.2 million individuals. In order to enable this passion for travel, many of the world's major airlines operate multiple daily flights into and out of Israel, connecting directly to every major European and American city. With shifts in the global economy and the rise in Israeli-Asian business opportunities, airlines such as Cathay Pacific and Hainan Airlines have also added daily direct flights connecting Tel Aviv with cities such as Shanghai, Beijing and Hong Kong. In many ways, technology and air travel options are just now catching up to Israelis desire to interact with and explore all corners of the globe.

For the Chinese people, the main purpose of traveling abroad has traditionally been to pursue the best opportunities for education. Throughout the various waves of Chinese international travel in pursuit of an elite education, a famous Chinese proverb has always rung true: "Read ten thousand books, travel ten thousand miles" (读万卷书，行万里路). This phrase accurately defines the hardworking and diligent Chinese scholar. This tradition dates back to the mid 1900's, beginning with Rong Hong. Born in 1828, Hong was the first Chinese student offered the opportunity to study at Yale University in the US. After completing his degree with honors and excellence, Hong returned to China in 1855 with a desire to help other Chinese students study in the United States. And so the precedent was set for Chinese students aiming to receive their degrees from elite American universities. In 1872, following continued relationships between the Chinese Educational Mission and Yale University, a group of young Chinese students would board a ship for America. Among them among were the "Father of the

Railroad" known as Tianyou Zhan, the founder of Fu Dan University, and the first president of Qinghua University.

Following the founding of the People's Republic of China in 1949 was the second wave of international education-focused travel. New China needed talents with advanced knowledge in the sciences and technology. In order to facilitate the acquisition of these skills, the Chinese government provided funding and scholarships for students to study in Russia, France, Germany, UK and other places around the world. This framework enabled many Chinese academics to travel abroad in order to expand their knowledge and return to China to join the growing national class of scholars. In the early 1990s, China experienced an additional wave of international travel when a significant number of Chinese students received numerous graduate studies scholarships from countries such as the US. This created an extreme amount of competition at top international high schools, a trend that we still see today, with students at China's most prestigious international schools vying for a limited number of spots at America's top universities and research centers. The most recent wave of travel began around 2006 when more and more Chinese students chose to study abroad to complete both their high school education and undergraduate degrees. Today, this remains a strong trend and has resulted in some of the most highly motivated and successful scholars coming out of China. With the increase of middle-class population in China, more and more families who have the means to do so decide to send their children abroad with hopes for giving them a better, brighter future.

According to the Institute of International Education's 2016-2017 Annual Report, there were upwards of 350,000 Chinese students who pursued higher education opportunities in the US in the 2016-2017 academic year. New analysis from the Institute of International Education shows that most international students come from Asia and head to North America and Western Europe. Additionally, according to a report from the Ministry of Education, record high numbers of students in China leave the country to study abroad. It was estimated that there were more than 800,000 students from China studying around the world in 2017. This is nearly four times more than any other country. According to China Study Abroad Development Report 2016, more than four million students studied abroad between 1978-2015. China is clearly doing something right in encouraging students to seek out international study opportunities. It is recognizing that travel is part of the equation to achievement and opening new doors to innovation. By enabling its students to leverage education as a way to improve their education and explore other places in the world, China is betting on the knowledge and capabilities that these enriching international experiences will uncover among their international students.

Chinese travel and immigration to the US may be motivated by educational opportunities, but it is not limited to student's study abroad opportunities. According to the Migration Policy Institute, "Chinese immigrants are now the third-largest foreign-born group in the United States after Mexicans and Indians, numbering more than two million and comprising five percent of the overall immigrant population in 2013." Although this is just a fraction of the total Chinese population, it is an extremely large and significant number when considering other immigrant populations in the United States. Chinese families that with the hopes of creating future opportunities of students have been able leverage American stereotypes of Chinese students to their advantage. As mentioned in our chapter on risk-taking, Chinese-American students are widely viewed as extremely smart and are typically described as 'over-achievers'. Their work ethic and willingness to learn has given them a unique advantage in both school and business positions. More and more Chinese families are realizing that traveling to foreign lands such as the US to pursue new opportunities can be their way forward in creating a better, more prosperous life for themselves and their families.

While it may account for the majority of international travel, education opportunities are no longer the only reason people in China are seeking to travel abroad. Despite the fact that outbound Chinese tourists still represent only 10% of the country's total population, it appears that the globalization trends and interconnected nature of the world has also influenced Chinese international travel. According to the 2016 reports by China Tourism Academy (CTA) and the official tourism research institute of China National Tourism Administration (CNTA), the number of outbound tourism in China reached 122 million people, an increase from the 2015 figure of 117 million. Additionally, China has been the largest source of outbound tourism to Thailand, Japan, South Korea, Russia, Maldives, and the United Kingdom. The Chinese tourist market abroad shows that those who are able to travel are excited to explore other parts of the world and create a tremendous market for tourism in foreign countries. They are also very likely to bring their experiences back to China and encourage their family and friends to follow in their footsteps. We are confident that this trend will continue, both for group travel which accounts for approximately 40% of travelers, and for individual travel, which accounts for the other 60%.

The rise in travel may be a result of the changes to the Chinese Hokou system that, historically, severely limited the ability of Chinese people to travel both within and outside of China. Hukou is at its core an organizational tool that is used to manage the vast number of people that are part of the growing Chinese population. It was created and is followed as a way of promoting political stability and avoiding risk. Formed as a result of a need to regulate the number of workers who could travel from more poor and rural villages to cities, Hukou has evolved into a system of regulations that continues to hinder travel and create obstacles for those seeking to explore places outside of China. However, the rules of the Hukou system have been vastly reformed since its integration into Chinese culture and tradition. What was once off limits and restricted (for example, moving to a big city for a chance to create and innovate), has now opened up for those who can prove they have employment opportunities and are able to establish a residency of a least six months. Similarly, there has also been a change in recognizing potential in students from small rural towns. If they can achieve the necessary scores, a bright student from rural China now has the opportunity to apply for and attend a prestigious school in a big city. The personal obstacles an individual may face in order to actually leave their rural community and attend schools in the city are still very real. However, unlike with previous generations that may have been automatically barred, educational opportunities do now exist for particularly gifted and dedicated students from rural areas. It is becoming more common for such students to be admitted to their preferred course of study, regardless of their background. These small but significant changes will continue to influence the Chinese culture of travel, both nationally and internationally. We are hopeful that the trends seen today continue to grow and encourage the people of China to explore the world around them.

As we review the travel traditions of both the Jewish and Chinese peoples, we must consider the lessons we can learn from the ways in which these traditions have led to future innovation. In the case of Israel, there appears to be a clear correlation between a nation of citizens so set on exploring all of the world's countries and cultures and these individuals' ability to innovate and disrupt a wide range of industries. And so, we must ask: How do these significant travel experiences contribute to the fulfillment of Israelis' goals and dreams? How important is it for Chinese children to travel, to experience new cultures, and how does this lead to their success?

Likewise, as we examine traditions of Chinese students studying abroad, we have established that this is a win-win for both the students and China as a nation. China's best and brightest often study abroad and return to become leaders in the Chinese entrepreneurial economy. However, their experience abroad is often limited to educational opportunities. Which begs the question: how can we expand travel for Chinese students beyond study abroad programs such that their personal experiences as international travelers are one of personal enrichment and discovery? Below are some of the ways in which Chinese parents, educators and students can infuse existing and new travel opportunities with elements of Jewish travel traditions that cultivate curiosity and future innovation.

1) <u>The Immersive method for studying abroad</u>: According to a study by Columbia Business School professor Adam Galinsky, those who have lived abroad are more creative. Galinsky's research found that the more countries someone has lived in - not as a tourist, but as an engaged part of foreign society and environment – the more creative potential they will have in their work and in their achievements. Parallel personality studies conducted by Dr. Julia Zimmermann and Dr. Franz Neyer found that when comparing the personality development of German students who studied abroad to those who stayed in their native country, those who traveled were more open and more innovative. Not only that, those who experienced a different culture were more likely to embrace the unknown, seek out challenges, take on more responsibility and act independently with high self-esteem and trust in their actions. While Chinese parents, educators and students might already be 'sold' on all the reasons to study abroad, they tend to view the opportunity to do so mainly through the perspective of achieving academic excellence. Here are **5 Essential Habits** for a more immersive, and ultimately more successful study abroad experience:

   I. Be curious about the unknown – whether it be a new cuisine, type of music, or simply a neighborhood you haven't seen yet in the city you're studying in – get out there and embrace opportunities to explore!

II. Don't try to figure everything out by yourself! The most successful people know they need to learn from others. Get comfortable seeking help and asking questions.

III. Gain a sense of independence and self-confidence by understanding what it means to take care of yourself. From how to navigate the Metro to ordering a tea at the local shop, from managing your own budget to taking care of household chores – be your own most reliable resource.

IV. Engage with people with completely different backgrounds. A great way to both meet new friends and understand the culture you live in is by helping out in your community. Try leveraging your student body for unique volunteering opportunities.

V. Take on a new sport and stick with it. Whether you're on an advanced team or a beginner league, you'll gain grit when you find a new physical challenge and work your way to improve at it.

2) <u>Pursue enriching, non-academic programs abroad</u>: There are an increasing number of opportunities for students to travel abroad outside of academic programs in a way that will still enrich their educational experience. From entrepreneurial competitions to unique skill development program to volunteer opportunities, the world is full of ways to travel for limited periods of time, together with other like-minded students. Perhaps such programs can be an add-on prior to or following an existing commitment to study abroad. Alternatively, Chinese students can leverage a particular passion of theirs, such as playing an instrument, writing code, or playing sports, in order to participate in an international gathering of students with activities revolving around this skill. While these experiences might require an investment of time, effort and resources, they can provide students with unique opportunities to experience a new culture and

engage with a multicultural gathering of individuals. Much like study-abroad programs, these environments force students to step out of their comfort zones and cultivate a sense of independence.

3) <u>Travel to Cultivate Innovation and Passion</u>: The mind can be opened to new ways of thinking through international experiences. When we navigate the unknown, we can become more self-confident, creative, and innovative. Simply traveling as a foreigner can make all of the difference in how an individual learns to problem-solve and interact with his or her surroundings. Tasks as simple as asking for directions or buying water can turn into an adventure during travel. These new experiences are also a great way to develop one's passions and interests. Writers and thinkers from across all ages and cultures have always postulated that they must find inspiration for their work and would very often travel in search of it. From painters seeking the perfect sunset to architects studying the skylines of a foreign city, we are all greatly influenced by our environments. Passions and curiosity thrive when we challenge our senses with new sounds, smells, languages, tastes, sensations, and sights. Whether alone as an adult or together as a family, the Chinese people can aspire to prioritize international travel and infuse their lives with new adventures.

4) <u>Interact with the Locals</u>: For many Chinese students it can be quite challenging to adapt to new ways of learning, interacting and making friends. Many who were brought up as the only child in the household now find themselves negotiating with roommates and coping with interpersonal dynamics very different from the ones at home. These soft skills are of great importance and shouldn't be overlooked, particularly in our era of mass globalization and international collaboration. The ability to effectively communicate with people from diverse backgrounds and cultures may ultimately determine your student's competitive advantage in our dynamic global market. Whether studying abroad or traveling for pleasure,

Chinese parents and educators must encourage students to commit completely to living in their new environment and interacting with both locals and other foreigners. It is worthwhile to note that the cross-cultural experiences of the Jewish people were critical in enabling them to not only appreciate and uphold their own traditions but also acquire an openness to the many types of people, cultures and innovative ideas in the world around them.

# Chapter 13:
# Distinct Traits of Jewish Nobel Prize Winners

We began our analysis of the cultural infrastructure of innovation by exploring and examining the unprecedented number of Jewish winners of the Nobel Prize. We now return our focus to this subject in an effort to provide additional real-world examples of how the innovation-driven behaviors and perspectives discussed throughout the previous chapters helped cultivate the ingenuity of numerous Jewish Nobel Laureates. By learning more about the Jewish men and women that were able to become true innovators, we can see how the concepts and ideas presented in this book serve as a common thread connecting the diverse achievements of Jewish Nobel Laureates over the past 120 years. As we take a closer look at each of the following examples, we will highlight the most prominent innovation-related characteristic or behavior of each Nobel Laureate such that Chinese parents, educators and students can further understand and internalize what it takes to become a true innovator.

## Albert Einstein, The Refugee

Albert Einstein's developments in the field of physics earned him the Nobel Prize in 1921. It is hard to imagine the modern theoretical study of physics without his contributions. Einstein was born to Jewish parents in 1879 in the town of in Ulm, Germany. Despite his renowned brilliance, Einstein did not always excel in the strict academic framework of the schools in which he was enrolled. On his first attempt he was denied entrance to the Swiss Federal Polytechnic in Zurich and was forced to take courses at a local canton school to bolster the remainder of his skills beyond physics and mathematics. At the time he wrote of his disdain for his studies when "the spirit of learning and creative thought was lost in strict rote learning." After finally passing the exams, Einstein enrolled at Zurich Polytechnic at the age of 17. His somewhat tumultuous life continued after graduation when he was unable to find a teaching position and had to accept a position working in the Bern patent office. Einstein spent his evenings working on developing his theories and in 1905, what is dubbed his "miracle year", he rose to fame after publishing numerous physics papers including the one which detailed his work on E=MC^2. His theory of relativity followed, and he went on to be awarded the Nobel Prize in 1921. Although he would later become a refugee in the truest sense of the word, Einstein's nomadic life began much earlier. He had been uprooted as a child as his father pursued business opportunities around Europe and he was forced to acclimate to new schools and diverse environments on a seemingly continuous basis. It was during these years that Einstein learned multiple languages, interacted with students and school friends from international backgrounds, and eventually was invited to be a guest lecturer and professor at numerous schools of international distinction. He couldn't have known at the time, but in many ways these years of travel, learning new languages and being constantly forced to grapple with the unknown would prepare him for a lifetime of innovation and the ability to succeed against all odds.

Although Einstein was raised in a non-religious household, he identified as Jewish both culturally and nationally and was always an outspoken supporter of the Jewish people and their endeavors. In fact, it was

Einstein's immense popularity along with his continuous support of Jewish issues made him a target of the Nazi party as they came to power in the 1920's. Over the next two decades, his theories were denounced publicly by the Nazi party and his work cast aside and labeled "Jewish physics". A Nazi publication entitled "100 authors against Einstein" was penned by those who sought to devalue his work. Einstein's books were burned, a bounty was put on his head, and his face appeared on the cover of a magazine with the title "Not Yet Hanged". He had no choice but to flee Germany and seek refuge elsewhere. Despite having been invited to countless universities around the world, being awarded a multitude of prestigious honors and professorships, Einstein left Germany not knowing where he could call home.

His refugee status found him travelling through Belgium to England where he pleaded with Turkish politicians for refugee status. After some time, he was able to emigrate to the Unites States where he arrived with an invitation to become a professor at Princeton University. Einstein remained under refugee status until he gained US citizenship in the 1940s. Any hopes he may have had of returning to Germany were lost when the Nazis raided his home in Belgium, prompting him to formally give up his German citizenship by turning in his papers at the German Consulate in Antwerp. After he was provided refuge in the United States, Einstein went on to contribute to one of the greatest and most dangerous innovations of the 20th century - nuclear weapons. He eventually became staunchly opposed to the use of the weapon, but at the time, having witnessed the horror of the Nazi regime, he had persuaded the President of the United States to develop work on the bomb, fearing what would happen if Germany was able to develop it first.

Beyond his impressive body of work, there are myriad lessons that Chinese students can learn from examining Albert Einstein's life. While it's impossible to identify one specific turning point or experience that fueled his creativity and innovative spirit, we can understand how the many risks, life-or-death situations, and refugee experiences he all contributed to his successful career as an innovator and Nobel Laureate. Chinese students will luckily not have to experience these same threats to their existence, however they can seek to understand how such peril can translate into making bolder choices. Whether it be choosing what subject to study or what profession to pursue, Chinese students can think like a refugee – like Albert Einstein – and perhaps make the unconventional choice because it is where their passion lies. When they are unsure how to proceed or have failed unexpectedly the way Einstein did as a young student, Chinese students can remember what it means to start over with nothing but your perseverance to succeed. By thinking like someone who could lose everything at any moment – Chinese students can leverage the "refugee mentality" in order to find their way to success.

## Robert Aumann, the Debater

Robert Aumann received the 2005 Nobel Prize in Economics for his work on conflict and cooperation through game theory analysis. However, Aumann's innovative spirit and creativity in the field of economics began decades earlier. Born in Frankfurt, Germany, Aumann was born into the rise of the Nazi regime. He was fortunate to flee as a young child, escaping Germany with his family and arriving in the United States just two weeks before the horrible events of *Kristallnacht* in November 1938. It was at this time that the Nazi regime carried out a mass attack on the Jewish people in which they destroyed thousands of Jewish businesses, homes, hospitals, schools and places of worship. Otherwise known as "the night of broken glass," the name *Kristallnacht* comes from the shards of broken glass that littered the streets after the windows of Jewish-owned stores, buildings, and synagogues were smashed with sledgehammers. Many describe this event as a turning point for Jews in Nazi Germany and it is reasonable to believe that Aumann's fate would have been completely different had his family not escaped when they did.

Aumann's family was very religious (known in Judaism as Orthodox) and upon arriving in the US, they chose to enroll him in a Yeshiva - a Jewish parochial school for young boys. At the Yeshiva, Aumann not only completed the curriculum that would enable him to attend a university one day (e.g. math, reading, writing, sciences and arts), but he also spent hours studying and debating Jewish religious texts, the interpretations of Jewish Rabbis, and the history of the Jewish people. Aumann highly valued his early education and the way his upbringing invited him to question and explore the information he was presented. He went on to study both mathematics at the City College of New York and MIT. After meeting famed theorist John Nash, Aumann was inspired to study game theory. In parallel with his research he became a professor at both Hebrew University of Jerusalem in Israel and Stony Brook University in New York. The enormous impact of Aumann's early childhood educational experience later inspired him to publish a paper entitled "Risk Aversion in the Talmud", combining his specialty of economics with his life-long, in-depth questioning of the lessons and values found within Jewish religious texts.

By the time Aumann received the Nobel Prize he had gained international recognition for his studies and discoveries. Despite his many accomplishments, Aumann remained a humble family man, often commenting on how giving a powerful lecture was just as meaningful as studying a passage of Talmud with his eighteen-year-old granddaughter. When asked to impart just one lesson unto his children and grandchildren Aumann replied, "I think people should do in life what they like to do – not what they think makes the most money, not what their parents tell them, but what they like. If you like something then you'll do it well, and if you do it well, you will succeed. That's it."

In Aumann one can see a real passion for learning. He exemplifies the spirit of someone who is committed to the in-depth research and exploration required for creativity and innovation. Taking Aumann's life lessons to heart, Chinese students can understand how a life steeped in the methods of debate can truly lead to greatness. Being able to understand a variety of perspectives and intellectually spar with others can often be the key to opening up new opportunities. Debate is not just a format of interaction to be mastered but rather a way of cultivating ideas based on comprehensive knowledge. Likewise, we can see how interdisciplinary approaches such as Aumann's analysis of both ancient Jewish texts and modern economic theory truly enriches one's knowledge and creativity. Perhaps more than anything, Aumann's life story serves as an important reminder that regardless of where you come from, it is always in your power to dig deeper, pursue your passion and become an innovator in the field of your choice.

## Ralph Steinman, the Risk Taker

Ralph Steinman's Nobel Prize is one of both great triumph and great sadness. After growing up in a Jewish family in Montreal, Canada in the 1940's, Steinman pursued his passion for medicine and scientific research, earning his medical degree from Harvard Medical School in 1968. Soon after graduating, Steinman discovered and named dendritic cells, the "messengers" of the human immune system, while working as a postdoctoral fellow. Despite Steinman's efforts to prove otherwise, for many years this discovery was not viewed as a great breakthrough or success story. Despite challenges such as losing his research funding and being overlooked or dismissed by colleagues in his field, Steinman risked his entire career in order to continue developing his ideas. It was not until the early 1990s that experiments based on immunotherapy and vaccines really came to the forefront and Steinman's work was recognized for its brilliance.

Steinman worked for over 30 years as a doctor of immunology when he found out that he had pancreatic cancer and would likely die within a year. Steinman decided that he would risk using his own body in his own experiments. He created a protocol of disruptive experiments based on his studies in the field of cancer immunotherapy in hopes of improving his odds. Steinman went as far as to remove part of his tumor for scientific study and sent portions to technicians and scientists in laboratories across the world in his efforts to find a cure. And then, in late October 2011, after a lifetime devoted to improving the lives of patients suffering from disease, Steinman passed away. In a true twist of fate, only three days later the Nobel committee reached out to Steinman via an email that he would never read and a voicemail that he would never hear that he had won the Nobel Prize in Physiology & Medicine for his discovery of the dendritic cell and its role in adaptive immunity. While the Nobel Prize is not awarded posthumously, given that they were unaware of his passing, the committee decided to award the Nobel to Steinman's family. It is notable that just days before his passing, Steinman was interviewed about the possibility of him receiving a Nobel Prize for his work and he joked with his family, "I know I have got to hold out for that. They don't give it to you if you have passed away. I got to hold out for that."

Steinman is the epitome of what it means to take risks. He refused to give up and was willing to do whatever was necessary to pursue his goals. Even when his discoveries were dismissed as unimportant – he found a way to keep going. Even when he had to win over colleagues and prove to funding agencies that his work had merit – he didn't give up on his research. Even when it meant implementing unconventional therapies on his own body in hopes of prolonging his life – he kept coming up with new protocols to try. Even when he was nearing death and barely had the strength to go on – he kept on pursuing a cure. It is notable that while his colleagues can never truly know if his life was extended as a result of his self-tested innovative immunotherapy treatments, Steinman did outlast his six to twelve-month prognosis, living for over four and a half years after being diagnosed with pancreatic cancer.

Steinman never stopped searching and seeking, risking what he had to along the way. Despite it not saving his life, Steinman's commitment to choosing the road less traveled or the unconventional approach translated into his greatest achievement and his status as a Nobel Laureate. As Chinese students study Steinman's journey, they can appreciate the importance of being willing to take risks, great or small, in the pursuit of innovation. Only by taking risks can students open up the world of possibilities and subsequent potential achievements. Perhaps they will encounter failure along the way, but if Chinese students approach risk with the knowledge that failure will be a stepping stone towards their goals, they can proceed without fear. Steinman's unique life prove that when taking risks, failing forward and creativity come together, innovation is never out of reach.

## Lev Landau, the Traveler

Born in Baku, Azerbaijan in 1908 under what was then the Russian Empire, Lev Landau grew up with Jewish parents both working in esteemed fields of study. His father was an engineer and his mother a doctor. By the age of 13 Landau had already completed his high school curriculum, but his parents considered him too young to be sent to university. Instead, they enrolled him in the Baku Economical Technical School for one year before he went on to complete his studies at Baku State University. Landau contemplated pursuing a number of areas of study, including physics, mathematics, and chemistry, but rather than choose one he eventually enrolled in courses from across each of those departments. He was awarded his first degree just two years later. At the early age of 16, Landau left his home in Baku to begin a life of travel and research around the world.

In 1924 he arrived in Leningrad, the city that was considered the most prestigious for physics research in the Soviet Union. After completing his degree, Landau knew he wanted to think more globally and pursued an opportunity to travel to complete his PhD. He received a grant from both the Soviet Union and the Rockefeller Foundation Fellowship to travel to Germany, Denmark, the United Kingdom, and Austria. It was during these travels that he met the leading scientists of his era and was able to broaden his scope of understanding. Through this global perspective, he understood the role his research played amongst the scientific innovations being made around the world. Landau returned to Azerbaijan with a new vision for the future of theoretical physics and soon became the leader in this field within the Soviet Union. The ten volumes he co-authored about the "Course of Theoretical Physics" is still used today in graduate level physics courses worldwide. Landau worked his way up to becoming the head of the Theoretical Division at the Institute of Physical Problems in Moscow. For over 25 years he led a team of mathematicians and physicists who were supporting the Soviet development of both the atomic and hydrogen bombs. In addition to his work for the government, Landau continued his research on quantum mechanics, theories of superconductivity, and quantum electrodynamics. He was awarded the 1962 Nobel Prize in physics for his mathematical theory of superfluidity. By many accounts he regards his work abroad as the most influential in his discoveries and innovations. He regarded himself as a student of Niels Bohr after a visit to his Institute of Theoretical Physics.

From a very young age, Landau had the vision and motivation to leave his hometown behind in search of knowledge, academic challenges and new perspectives. By leveraging his international experiences, he was able to discover the global developments in physics that expanded far beyond what he had been able to learn in his native Azerbaijan. While many Russian scientists felt the need to prove their loyalty to the Soviet Union by remaining Moscow and St. Petersburg, Landau had the foresight to understand that the only way to meet the leading innovators in his field was through travel. Lev Landau was rewarded for his adventurous spirit and was admitted as a foreign member of various prestigious academies like the Danish Royal Academy of Sciences, Royal Society of London and Netherlands Royal Academy of Sciences. He also became an honorary member of the American Academy of Arts and Sciences and of the Physical Society of France. Additionally, Landau served as the Foreign Associate at the National Academy of Sciences of the United States.

Chinese students have much to learn from Lev Landau and his travels across the globe. It takes courage to decide to leave the comfort of home. And yet, if Chinese students are able to show courage by stepping out into the world and exploring what it has to offer, they will expand their minds and gain greater insight on how to realize their dreams. Only by comparing their skills, talents and achievements to those of people around the world can students truly understand what they are truly capable of. As Mozart once said, "A man of ordinary talent will always be ordinary, whether he travels or not; but a man of superior talent will go to pieces if he remains forever in the same place." As we reflect on the accomplishments of Lev Landau, we can see that only by studying and immersing oneself in other cultures can an individual discover his or her true place in the world.

## Rita Levi-Montalcini and her Chutzpah

Born in Turin, Italy in 1909, Rita Levi-Montalcini was a woman who defied all the odds. Not only was she limited by the archaic views on women's education held by the patriarchal and Victorian society she grew up in, but as a Jew she was barred from academic and professional careers following Mussolini's Manifesto della Razza. One of her first obstacles was convincing her father to let her leave her girls' high school and abandon the traditional path of marriage at age eighteen. With great hesitation and fear for his daughter's future, Levi-Montalcini's father agreed to let her pursue her dream of studying medicine at Turin School of Medicine. She passed her entrance exam and enrolled at age 21. Levi-Montalcini excelled in her studies and quickly gained recognition as an intern at the Institute of Anatomy where she conducted research on the nervous system of chick embryos.

In the 1930s, under the fascist regime of Mussolini, hostility towards the Jewish people of Italy reached its peak. Levi-Montalcini was forced to withdraw from her university studies as to avoid risking both her safety and the safety of her non-Jewish colleagues who would be implicated were she to be found studying among them. Despite this enormous setback, Levi-Montalcini remained undeterred and instead created a laboratory in her bedroom. In the years that followed, every time bombs would fall nearby, she brought her microscope and slides to safety in the basement of her home. In 1943 she was forced into hiding in Florence and remained there until the end of World War II. This was yet another challenge that she would conquer, and throughout the war she continued her research and writing in secret. Levi -Montalcini eventually went on to accept a professorship at Washington University in St. Louis, where she was able to complete the important work that would eventually lead to her Nobel win for her discoveries in the field of isolating nerve growth factors. After decades of research, at age 77, Levi-Montalcini received the Nobel Prize in Medicine. While she was proud to receive this tremendous honor and recognition of her work, Levi-Montalcini continued her research at her apartment in Rome, living until the age of 103.

Levi-Montalcini ascribes her success to the adversity and hardships she had to face throughout her life. She had the chutzpah, the daring, the gall, and the audacity to reject what was traditionally expected of her and pursue her career as a woman in science. As a Jew during WWII, she was constantly discriminated against and, at the height of the war, she often feared for her life. Levi-Montalcini spent her whole life challenging stereotypes and proving to everyone around her what she was truly capable of. It was the multitude of challenges she faced along the way that made Levi-Montalcini's Nobel Prize win all the more momentous.

She learned to embrace the forces that shaped her, declaring, "If I had not been discriminated against or had not suffered persecution, I would never have received the Nobel Prize".

Levi-Montalcini provides Chinese students, particularly young women, with an incredible example of what it means to use chutzpah, boldness and audacity to succeed. She knew her worth even when everyone else questioned it and was willing to put her life on the line to pursue her passion for medicine. When an individual understands his or her worth and what he or she is capable, nothing can get in the way of success. If students today can begin cultivating their chutzpah and making daring choices, they are that much more likely to realize their potential for innovation. Much like Levi-Montalcini and many other Jewish Nobel Laureates, this journey must begin with each student gaining a fundamental understanding of what he or she capable of – only then can chutzpah be used to transform dreams into reality.

## Henry Kissinger, Speaker of Languages

Heinz Alfred Kissinger was born in Fürth, Germany, in 1923. Raised in an Orthodox Jewish household, Kissinger studied the Bible and the Talmud in Hebrew as part of his daily routine. From a young age Kissinger felt the rise of anti-Semitism around him. He excelled as a student and had hoped to be accepted to the notable state-run Gymnasium, but by the time he could apply, Jews were forbidden to study there. In 1938 the Kissinger family decided to escape the ever-worsening reality Jews faced in Germany and set off for New York City via London. Despite having achieved significant economic success in Germany, the Kissinger family arrived in the United States as refugees with nothing to their name.

Kissinger immediately went to work in a shaving brush factory in order to help contribute to the household income. Heinz changed his name to

Henry and began studying at George Washington High School. He excelled in all of his classes and learned English especially quickly. One teacher recalls that, "He was the most serious and mature of the German refugee students, and I think those students were more serious than our own." Kissinger graduated from high school and studied to become an accountant at the City College of New York. It was shortly after this that the United States entered World War II and Kissinger was drafted into the military. He served first as a rifleman in France but was then moved to Germany to work as a G-2 Intelligence officer. It was there that Kissinger says he felt the most American.

He was now using one of his native languages, to help the US defeat the regime of his native Germany. He was later tasked with leading US Army teams that brought civilian administration and order back to German towns which were ravaged and left in ruins by the Nazi party. After completing his military service, Kissinger returned to the United States to pursue his Bachelor's degree at Harvard. Remaining at Harvard for his Ph.D. studies, Kissinger was later asked to join the faculty in the Department of Government. Although he had been completely immersed in the world of academia at the time, Kissinger had always been interested in a career in government. He began his career in Washington, D.C. as a trusted advisor to President John Kennedy in 1961, the first of many US Presidents that would include Kissinger in their core team. President Richard Nixon appointed Kissinger Secretary of State where most of his political workings in the field of 'shuttle diplomacy' brought stability to numerous nations of the Middle East. He was awarded the Nobel Peace Prize in 1973 for his involvement in negotiating a ceasefire in Vietnam.

Kissinger is often seen as a controversial figure and both his career and his political views have been the source of much debate. However, even his most adamant adversaries would agree that his language skills were extremely impressive and they played a critical role throughout his life.

From learning English quickly in order to excel in the US education system to leveraging his fluency in German to move up through the ranks of the US Army, Kissinger was able to use his multilingual abilities to achieve success across different sectors. By realizing the importance of learning a second (or even a third) language at an early age, Chinese students can too use this to their advantage in their careers. During Kissinger's lifetime, his ability to speak both German and English was highly valued. In today's world, one could argue that learning Robotish, or the language of code, will be the most valuable for succeeding in the future. Our global society is quickly moving towards a reality in which IT, technology and computer sciences will be integrated into all aspects of life. As a result, whatever subjects students choose to study or whatever professions they pursue, learning Robotish will enhance of their future career and enable innovation across all fields of study.

## Yitzhak Rabin, the Military Man

Born in Jerusalem in 1922, Yitzhak Rabin was a great statesman, military man and diplomat. He is considered one of the early pioneers and founders of the State of Israel and played a critical role in its development. At the age of 19 Rabin had already been recruited to join the *Haganah*, a Jewish paramilitary organization in the British Mandate of Palestine (1921–48), which became the core of the Israel Defense Forces (IDF). That same year he became first to enlist in the *Palmach*, an elite fighting force of the *Haganah*, and became deputy commander of the First Battalion while still in his early twenties. In 1947, after numerous battalion activities throughout the country, he became the *Palmach*'s Chief Operations Officer.

His military career also included being sent to England to study at the British army's Staff College. Rabin was renowned for his ability to both think quickly when necessary and develop excellent strategies whenever possible. In the War of Independence in May of 1948, Rabin led the forces that were able to successfully reclaim Jerusalem for the Jewish people. As the Palmach's Operations Officer, he planned all major wartime strategies and was even sent as a representative to the first ceasefire talks with the Egyptians. When the IDF was eventually founded, Rabin was called upon to integrate the Palmach and all other military training branches in order to establish the official IDF command and set the standards for officer training.

Rabin's impressive career of service and leadership continued to play an important role in the establishment and growth of the state of Israel. He went on to serve as Ambassador to the United States, proving that he could excel not only as a military leader but as a diplomat charged with what was considered one of the most critical international partnership Israel needed to cultivate. After successfully rising through the ranks of both the military and the government, Rabin eventually became Prime Minister of Israeli in 1974. His two terms as Prime Minister can be described as the source of real change and leadership. His leveraged both his military experience and political prowess to in his efforts to establish peach with Israel's neighbors in the Middles East. In 1994 he was awarded the Nobel Peace Prize for his continuing efforts to continue peace talks through the Oslo Accords. Rabin experienced a great deal before the age of thirty. He commanded troops of battalions. He had to make quick decisions to keep his soldiers safe and at the same time, had to keep in mind the long-term, important goal of one day establishing an independent state of Israel. Rabin's military career is quite inspirational and truly changed the future for a nation filled with Jewish WWII refugees. He was able to use all of his acumen to put the young state of Israel on the international map. These same abilities would later drive in him as a politician and leader of a nation.

Chinese students have much to learn from this impressive career of military and national service. Rabin was successful not because he simply followed orders but because he was able to think quickly and decisively even when under enormous pressure. Students today can use Rabin's example to understand that the great respect they have for their teachers and mentors need not mean following their every request but rather finding opportunities for free thinking and ingenuity. Likewise, educators and parents can assist students in laying the groundwork for the creative and critical thinking that leads to the cultivation of an innovative mind. Much the way Rabin did with his soldiers, parents and teachers must not only prepare their students for the world, but then be willing to put their complete trust in their students in order to truly let them fly on their own. In doing so, students will feel confident that whatever mistakes they may make or challenges they may face – they are capable of coming out on top. These types of leadership skills are common in military heroes like Rabin and can be used to improve the lives and futures of students who will grow into tomorrow's leaders.

## Gertrude Elion, Creator of Balagan (Chaos)

Born in New York City in 1918 to Jewish immigrant parents from Poland and Lithuania, Gertrude Elion had a natural curiosity and thirst for knowledge that would shape her entire life. And yet, like many women of her era, she was faced with a series of seemingly insurmountable obstacles on her path to success. As a young school child, Elion remembers being not being able to decide what to study after high school but eventually decided to pursue her love of science. She earned a bachelor's degree in Chemistry from Hunter College and a Master of Science from New York University. However, all fifteen of her fellowship applications were denied, simply because she was a woman. She was forced to pursue a new direction decided to attend secretarial school for six weeks, a profession where women were allowed to work at the time. Following the course, she ended up working as a food quality supervisor at a supermarket, testing everything from the acidity of pickles to the color of egg yolk in mayonnaise. Still somewhat unsatisfied with this position, Elion sought out another way to pursue her passion for science. She managed to enroll in part-time doctorate program at the Brooklyn Polytechnic Institute, only to be later told that she couldn't work and simultaneously pursue her studies, and due to the barriers against woman in the sciences, she was never granted the opportunity to finish her PhD.

As luck would have it, the chaos of World War II would translate into opportunity for women like Elion. When the war broke out, many positions opened in industrial laboratories that needed chemists. Since the men were drafted into the military and sent off to war, any women with the right qualifications were recruited into technical positions that had been customarily filled by men. Elion was accepted to the Hitchings lab and became involved in microbiology. She leveraged the fact that she could now develop her chemistry research and continued experimenting while broadening her horizons into the fields of biochemistry, pharmacology, immunology, and virology. Elion's interests expanded far past her work as a scientist. She was an avid photographer, traveler, and Metropolitan Opera subscriber for over 40 years. Her life may have seemed chaotic from the outsider's perspective, a dynamic constellation of all of her many interests, fields of study and professional roles. However, this same passion for exploration across different areas would eventually lead to her becoming a Nobel Laureate. In 1988 she received the Nobel Prize for "important new principles of drug treatment" including her work in experimental therapy, and the adaptation of AZT- the first drug used for the treatment of AIDS.

Elion's early studies and varied job experiences exemplified the word *balagan*. An objective onlooker might look at her resume and instead of seeing the career path of a brilliant scientist they would see a lost soul scattered in too many directions. But for Elion, dabbling in many different fields was not about being disorganized, it was about overcoming challenges and finding her way even when she was told 'no.' She always realized that her ultimate goal was to work on cancer research, but she was able to learn, discover new ideas and leverage a wide variety of experiences along her 'chaotic' and roundabout path in order to ultimately reach her goal. Elion's experience shows us that things are not always as they seem. More importantly, it shows us that taking chances and going against the grain can eventually put you at the right place at the right time. Balagan, and the outwardly appearing disorder of a life filled with many passions, can actually create order and interdisciplinary connections in the mind. As an explorer, scientific researcher and lover of art, music and photography, Elion was committed to a life of discovery. Young Chinese students of today may follow Elion's example by pursuing their passions across numerous fields. This might require going against the gain, but ultimately it will enrich all of their studies and experiences. Expanding students' interests by committing to all extracurricular activities in a passionate, meaningful way can be key to unlocking the innovative ideas within us. This exploration and discovery process, even when it may seem chaotic, can ultimately provide students with the opportunity to succeed, just as it did for Elion.

# Chapter 14: Conclusions

As we have seen throughout our examination of various aspects of Jewish and Chinese culture, these cultures share more in common than it might appear at first glance. From concepts taken from Jewish religious texts and Chinese axioms such as the Golden Rule, we can see that there is a great deal of overlap and intersection between the underlying values and beliefs upon which both cultures are built. Much in the way that Confucian teachings and tenets of Taoism shape Chinese traditions of family, hospitality and empathy, Jewish Rabbis' interpretations of religious teachings influenced these same traditions among the Jewish people, resulting in a myriad of cultural norms among both peoples that stem from similar ideas and customs. It is through our high regard and respect for these elements of commonality that we were able to extrapolate from these basic shared values in order to explore the Jewish cultural infrastructure of innovation. We did so in search of lessons that both apply to and enhance the innovative capabilities of students throughout China.

Throughout the past 13 chapters of this book, we have journeyed through a vast array of historical and modern-day approaches to innovation among the Jewish people as well as informative examples from the lives of successful Jewish innovators. In an effort to truly prepare Chinese parents, educators and students for a life of innovation, we will now zoom out and examine the overarching themes that appear to play a particularly central role in the process of cultivating creativity. By presenting the axioms that span across multiple aspects of innovation, we can show the underlying connections between them. Furthermore, we can demonstrate how concurrently implementing many of the recommended lessons and practices will create the most impact in the lives of Chinese students.

The first and perhaps most apparent of these truisms is that there is no one path to innovation and there are certainly no shortcuts. The constant challenges the Jewish people faced throughout history forever changed the lens through which they view themselves and our global society. However, by overcoming a seemingly never-ending series of fatal threats, the Jewish people proved both to themselves and the world around them that adversity and dealing with the unexpected are inherently part of life. The changes, risks and adventures that result from such challenges are not something to be feared or a reason to give up, but rather can be used as fuel to achieve even the seemingly impossible. It is therefore no surprise that many of our lessons focus on learning how to fail, taking risks and challenging yourself by stepping outside your comfort zone. The greatest of business men and women to this day have made risk taking part of their daily routine and we must encourage our students to do the same. How else will we find the next Jack Ma, Sheryl Sandberg or Elon Musk? Where will we discover the next Liu Chuanzhi, Mark Zuckerberg, Robin Li, or Cher Wang?

Whether it be through military training or acting with audacity, Chinese students can use the risk-taking opportunities they create for themselves to truly test their limits and see what they might be capable of when they travel the path unknown. They might be greeted with chaos and uncertainty. It might require them to go against the grain and leave their home behind in search of new experiences and cultures. It might mean that they spend more time cultivating 21$^{st}$ century skills like Robotish in lieu of more traditional rote learning tactics. Wherever these adventures may take them, if Chinese students can take on the world with a little bit of chutzpah, they will surely come out on top.

The second axiom that is woven throughout the previous chapters is the importance of cultivating both passion and grit. From our analysis of new educational methodologies to the constant need for asking questions and being curious, to our recommendations for creating the optimal study abroad experience, we have clearly established that that exploring and asking questions as a way to find and cultivate one's passions is for enabling future innovation. Likewise, these passions are almost meaningless if students don't concurrently develop their ability to follow through and pursue their interests with vigor, determination and courage. As you may have noticed, the themes of asking questions, delving deeper into ideas and pursuing one's interests professionally appears in various forms throughout all of the previous chapters.

That said, we realize that in many ways such recommendations and behaviors are antithetical to traditional Chinese approaches to learning. However, we must remember that no amount of practice can replace the need for a student to have an inherent interest in the particular skill or subject they are pursuing. Students must be encouraged to discover and explore what makes them 'tick' in order to find the path that will truly fulfill them - rather than the one that might "look" the best to peers, colleagues or admissions offices. Likewise, students that are able to develop stamina and create long-terms goals are able to internalize Angela Lee Duckworth's concept of "grit". Grit does not measure talent but rather a student's reaction to short-term failure and the continued perseverance of a long-term goal. The innovators we admire most as a global society consistently failed forward and did so in pursuit of excelling in the field they were passionate about.

Similarly, there is a need to create environments that allow these processes to happen naturally and without fear or pressure. Chinese students can feel paralyzed by fear of making a wrong choice and are many times focused solely on the test in front of them rather than how they feel about what they are studying or what they would want to learn more about within their vast curriculum. By enabling our students to see that their education and subsequent career is a marathon rather than a series of short sprints, we can give them the mental and physical space to pursue a new skill or idea without it coming at the cost of what they feel they might be required to do. The passions students develop as children are often what can shape future creativity, ingenuity, innovation and success. Students who have both passion and grit can use their studies to find a path that is not only practical but also meaningful to them. Ultimately, these students will be much more likely to bring innovation to what they do and contribute significantly to the world around them.

The third and final conclusion we wish to leave you with is the one that is perhaps the most difficult to internalize. It requires that Chinese parents and educators to **embody the change they wish to see in their students**. While we may have thousands of years of history filled to the brink with role models to learn from and from which we can seek inspiration, there is nothing that can match the influence or impact parents and educators have on their students. If we wish for our students to become more innovative, we must be willing to serve as the modern, every day role models that they can emulate in order to learn and adopt the characteristics and behaviors found in innovators. Perhaps some will find it easier to model what it means to ask great questions. Others will feel more comfortable showing their students what it means to take risks. Regardless of the methods we choose, Chinese parents and educators must work together with their students, side by side, to not only teach but demonstrate the lessons provided throughout this book. In doing so, we can successfully cultivate the next generation of innovators and Nobel Laureates.

Final Thoughts and Recommendations

Through our extensive examination of historical, religious and cultural examples, this book has sought to illuminate the strong connection between the way the Jewish people experience life and their ability to successfully innovate across diverse sectors and disciplines. From the elements of innovation among the Jewish people consciously passed down through generations to those that are learned more passively within Jewish society, they can be used to inspire our students to aspire to greatness. Through our exploration of essential character traits, traditions and perspectives, we hope this book was able to provide a roadmap to innovation for the Chinese students of today and for generations to come. As you set off in pursuit of ingenuity, we leave you with our concise, curated list of "10 commandments" that reflect the main lessons found in this book.

**The 10 Commandments of Success, Ingenuity and Innovation:**

1. Think Like a Refugee: Take a chance TODAY, we can never know what tomorrow will bring.
2. Ask Questions & Debate to Find the Right Answers: New ideas and innovative solutions can only be found by asking questions. Use debate to discover how differing perspectives can bring you closer to the truth.
3. Take Risks & Act Boldly: Take a leap of faith. If you are afraid to fail, you may never reach your full potential.
4. Travel & Study Abroad: Open your heart and mind by exploring a foreign country and culture. Set off to find yourself in the world and return home a better version of yourself.
5. Show Some "Chutzpah": Tap into your full potential by audaciously pursuing your dreams and not letting anything stand in your way.
6. March to the Beat of Your Own Drum: Follow your passion and develop your own sense of self-awareness and self-confidence.

7. <u>Embrace the "Balagan"</u>: A little chaos is a necessary part of the innovation process. Stay the course and you'll find there's something wonderful just around the corner!
8. <u>Be Multi-Lingual</u>: Global communication can only occur when we all understand each other and are speaking the same language.
9. <u>Coding is Key</u>: It's the 1's and 0's that will make or break the next great innovative mind.
10. <u>Be Best and Be First</u>: True ingenuity can be found at the intersection of a creative spark and an unstoppable spirit. Find a way to combine your passion with a commitment to the fail-forward, entrepreneurial spirit that will keep you one-step ahead of the competition.

# Author Biographies

**Ami Dror** is the founder & CEO of LeapLearner, an education-technology company that empowers students through the acquisition of 21st century skills such as coding and problem-solving. Based in Shanghai, China, LeapLearner's platform gives students around the world the tools to leverage their innovation and creativity in order to code the future. Ami is also the co-founder of Zaitoun Ventures, a values-driven startup factory that that partners with diverse global partners in order to build disruptive, successful startups across multiple sectors.Prior to founding Zaitoun Ventures, Ami had a successful public-sector career as head of security for the Israeli Prime-Minister and as Vice -Consul in Israel's Ministry of Foreign Affairs. Moving into the private sector, Ami co-founded XPAND 3D, and led its growth into a global 3D Cinema industry frontrunner. Additionally, he initiated the work on the first 3D-TV and led the formation of a 3DTV licensing company, jointly owned by Sony, Samsung, XPAND and Panasonic. Ami played an integral role in the development of *Amblyz* that revolutionized the treatment of childhood amblyopia. Ami serves as a member of the board of directors of several companies. He holds a BSC in computer engineering from the Open University in Israel and is the co-founder of China Israel Innovation fellowship, a Hong Kong based NGO that promotes global innovation. Ami is a Henry Crown Fellow at the Aspen Institute and an active member of the Aspen Global Leadership Network.

**Dr. Jordan (黄兆旦) Zhaodan Huang** is the founder of Uni-Ed, a cross-border educational consulting firm aiming to provide Chinese students with a one-stop shop for education solutions. In addition to offering college counseling guidance, follow-up service in the US and post-graduation employment guidance, Uni-Ed also organizes customized summer and winter programs for leading Chinese middle schools, high schools and universities. After Uni-Ed was acquired by Enoch Education in 2018, Huang joined the Board of Directors where she manages the company's family education consultancy practice and oversees corporate marketing and branding. Prior to founding Uni-Ed, Ms. Huang served as the Vice President of Teneo, a consulting firm advises the world's largest organizations with holistic advisory solutions. At Teneo she leveraged her expertise in order to advise Chinese companies on their overseas development opportunities and American companies on their PR strategy and operations in China. Prior to Teneo, Ms. Huang worked as the Chief Academic Officer and VP Business Development of ActiveChinese, an educational multimedia software and online content provider of Mandarin Chinese for clients ranging from the United Nations to US public school districts. Ms. Huang also previously served as Principal of the Chinese Language School of Connecticut (CLSC) where she managed over 800 students and staff and helped provide Chinese language and cultural consulting services to NBC, BBC and many private schools. Prior to CLSC, Ms. Huang served as the Chinese Language Program Director at China Institute, the oldest bicultural organization in America devoted exclusively to China. During her tenure at China Institute, she worked with Nickelodeon, Sesame Workshop and NY1 on introducing Chinese language and culture to broader U.S. audience. Prior to that, she was a part-time evening News Anchor for Sinovision Inc., the largest Chinese TV news and program provider across U.S. Ms. Huang obtained both her B.A. and M.A. from Peking University and her Doctor of Education degree in International Educational Development from Columbia University.

# References

"A History of the Rothschild Family." *Investopedia*, 23 Feb. 2017. Accessed Apr-May 2017.

Alban, Deane. "The Brain Benefits of Learning a Second Language." *Be Brain Fit & Blue Sage, LLC*, n.d. https://bebrainfit.com/brain-benefits-learning-second-language/. Accessed Apr-May 2017.

Angle, Stephen C. "Social and Political Thought in Chinese Philosophy." *Stanford Encyclopedia of Philosophy*, 21 Jul. 2016. https://plato.stanford.edu/entries/chinese-social-political/. Accessed Apr-May 2017.

Arieli, Inbal. "How Israeli Culture Promotes Creativity and Independence." *Jewish Journal*, 3 May 2017. http://jewishjournal.com/news/israel/218634/israeli-culture-promotes-creativity-independence/. Accessed Apr-May 2017.

Aumann, Robert J. "Robert J. Aumann – Biographical." *Nobelprize.org*. Nobel Media, 27 Jun. 2017. http://www.nobelprize.org/nobel_prizes/economic-sciences/laureates/2005/aumann-bio.html. Accessed Apr-May 2017.

Botwinick, Chaim. "Leadership and Followership in the Judaic Classroom: Challenges and Opportunities." *The Lookstein Center for Jewish Education*, vol. 15, no. 1 (Winter 2016), https://www.lookstein.org/journal/leadership-followership-judaic-classroom-challenges-opportunities/. Accessed Apr-May 2017.

Brown, Laura Lewis. "Comparing Preschool Philosophies: Montessori, Waldorf and More." *PBS Parents*, n.d.
http://www.pbs.org/parents/education/going-to-school/choosing/comparing-preschool-philosophies-montessori-waldorf-and-more/.
AccessedApr-May 2017.

Ebert II, Edward S., Christine Ebert, and Michael L. Bentley. "Methods of Teaching in the Classroom."
https://www.education.com/reference/article/methods-teaching-classroom/.
From *The Educator's Field Guide*. Corwin, 2011.
Accessed Apr-May 2017.

Epstein, Varda. "The Debate over Jewish Achievement: A Review." *The Huffington Post*, 25 Aug. 2015.
http://www.huffingtonpost.com/varda-epstein/the-debate-over-jewish-achievement_b_8027028.html.
Accessed Apr-May 2017.

Gladwell, Malcolm. *The Tipping Point: How Little Things Make a Big Difference*. Little Brown, 2000.

Gunn, Dwyer. "How Did Israel Become 'Start-Up Nation'?" *Freakonomics*, 4 Dec. 2009.
http://freakonomics.com/2009/12/04/how-did-israel-become-start-up-nation/.
Accessed Apr-May 2017.

Halpern, Micah D. "The Art of Debate: Jewish Style." *Asia Society*, n.d.
http://asiasociety.org/countries/religions-philosophies/art-debate-jewish-style.
Accessed Apr-May 2017.

Hamza, Khalid and Bassem Alhalabi. "Technology and Education: Between Chaos and Order."
*First Monday: Peer-Reviewed Journal on the Internet*, vol. 4, no. 3-1, Mar. 1999,
http://firstmonday.org/ojs/index.php/fm/article/view/656/571.
Accessed Apr-May 2017.

Hariharan, Anu. "On Growing: 7 Lessons from the Story of WeChat." *YCombinator.com*, 12 Apr. 2017.
https://blog.ycombinator.com/lessons-from-wechat/.
Accessed Apr-May 2017.

Holloway, Marguerite. "Finding the Good in the Bad: A Profile of Rita Levi-Montalcini." *Scientific American*, 30 Dec. 2012.
https://www.scientificamerican.com/article/finding-the-good-rita-levi-montalcini/.
Accessed Apr-May 2017.

Hooda, Saurabh. "3 Reasons Why Everyone Should Learn Programming." *Entrepreneur India*, 16 Feb. 2017.
https://www.entrepreneur.com/article/289248.
Accessed Apr-May 2017.

"How the Rothschilds Created Modern Finance and a Vast Fortune That Has Lasted for Centuries." *Business Insider*, n.d.
http://www.businessinsider.com/the-early-rothschilds-built-a-fortune-2012-12?op=1/#e-rothchilds-come-from-humble-beginnings-the-jewish-ghetto-in-frankfurt-known-as-the-judengasse-1.
Accessed Apr-May 2017.

Institute of International Education. 2017 "Top 25 Places of Origin of International Students, 2015/16-2016/17. Open Doors Report on International Educational Exchange. Retrieved from https://www.iie.org/Open-Doors/

"Is Having Grit the Key to Success?" *NPR*, 1 Nov. 2013.
http://www.npr.org/templates/transcript/transcript.php?storyId=240779578.
Accessed Apr-May 2017.

Jewish Women's Archive. "Gertrude Elion." (Viewed Apr-May 2017) https://jwa.org/womenofvalor/elion.

Jewish Women's Archive. "Rita Levi-Montalcini Wins the Nobel Prize." (Viewed Apr-May 2017) https://jwa.org/thisweek/oct/13/1986/rita-levi-montalcini.

Kaplan, Jonathan. "The Mass Migration of the 1950s." *The Jewish Agency*, 27 Apr. 2015.
http://www.jewishagency.org/society-and-politics/content/36566.
Accessed Apr-May 2017.

Kent, Orit. "A Theory of *Havruta* Learning." The Beit Midrash Research Project: A Project of the Mandel Center for Studies in Jewish Education, Brandeis University.
https://www.brandeis.edu/mandel/pdfs/TheoryofHavrutaLearning.pdf.
Accessed Apr - May 2017.

Kerbs, Gil. "'It's Not Rudeness, It's Chutzpah' – An Insider's Take on Israel's Startup Success." *The Heureka*, 22 Oct. 2012.
http://theheureka.com/gil-kerbs-chutzpah-israel-startups.
Accessed Apr-May 2017

Kopin, Brett. "Voices from the Field: An Interview with Nobel Prize Winner Robert Aumann." The iCenter, 1 Aug. 2015.
http://www.theicenter.org/voice/interview-nobel-prize-winner-robert-aumann.
Accessed Apr-May 2017.

Lederhendler, Eli. "Orphans and Prodigies: Rediscovering Young Jewish Immigrant 'Marginals.'" *Project Muse*, vol. 95, no. 2,
http://muse.jhu.edu/article/399288.
Accessed Apr-May 2017.

Levitt, Steven D. and Stephen J. Dubner. *Freakonomics: A Rogue Economist Explores the Hidden Side of Everything*. William Morrow, 2005.

Lynn, Richard. *The Chosen People: A Study of Jewish Intelligence and Achievement.* Washington Summit Publishers. 2011.

Mitgang, Herbert. "48 Jewish Nobel Laureats Honored." *The New York Times*, 13 Nov. 1986.
http://www.nytimes.com/1986/11/13/books/48-jewish-nobel-laureats-honored.html?mcubz=1.
Accessed Apr-May 2017.

Mitzner, Dennis. "5 Reasons Behind Israel's Startup Success." *The Next Web*, 7 Jul. 2015.
https://thenextweb.com/insider/2015/07/07/5-reasons-behind-israels-startup-success/#.tnw_jRAs5BF8.
Accessed Apr-May 2017.

Morris, Ruth. "Parliamentary-style Debates Take Off in China – Even If Some Topics Are Off Limits." *Pri.org*, 5 Sept. 2014.
https://www.pri.org/stories/2014-09-05/some-topics-are-limits-parliamentary-style-debates-are-becoming-popular-china.
Accessed Apr-May 2017.

"Nobel Prize Facts." *Nobelprize.org*, 27 Jun. 2017.
http://www.nobelprize.org/nobel_prizes/facts/
Accessed Apr-May 2017.

Parker, Barry R., ed. *Einstein: The Passions of a Scientist.* Prometheus Books, 2003.

Pease, Stephen L. *The Debate Over Jewish Achievement: Exploring the Nature and Nurture of Human Accomplishment.* Deucalion, 2015.

Pew Research Center. "Religion and Education Around the World." 13 Dec 2016.
http://www.pewforum.org/2016/12/13/jewish-educational-attainment/

Pew Research Center. "How income varies among U.S. religious groups." 11 Oct 2016.
http://www.pewresearch.org/fact-tank/2016/10/11/how-income-varies-among-u-s-religious-groups/

Pevzner, Alexander B. "Op-Ed Israel-China Romance is Based on Ancient Values." *JTA.org*, 30 Sept. 2015.
http://www.jta.org/2015/09/30/news-opinion/politics/op-ed-israel-china-romance-is-based-on-ancient-values.
Accessed Apr-May 2017.

Pontz, Zach. "Richard Dawkins Perplexed by High Number of Jewish Nobel Prize Winners."
*The Algemeiner*, 29 Oct. 2013.
https://www.algemeiner.com/2013/10/29/richard-dawkins-perplexed-by-high-number-of-jewish-nobel-prize-winners/.
Accessed Apr-May 2017.

Razeghi, Andrew., ed. *The Riddle: Where Ideas Come From and How to Have Better Ones*. Jossey-Bass, 2008.

Reshef, Tal. "Israeli Chutzpah a Hot Commodity in China." *YNetNews.com*, 10 Sept. 2010.
http://www.ynetnews.com/articles/0,7340,L-3963443,00.html.
Accessed Apr-May 2017.

Rogers, Everett M., ed. *Diffusion of Innovations, Fifth Edition*. Free Press, 2003.

Romberg, Jack. "A Tradition of Questions, Questioning Tradition." *The Jewish Observer*, 5 Feb. 2013.
https://thejewishobserver.com/2013/02/05/a-tradition-of-questions-questioning-tradition
Accessed Apr-May 2017.

Schilpp, Paul Arthur, ed. *Albert Einstein: Philosopher-Scientist, Volume II*. Harper Torchbooks, 1959.

Schulte, Mark. "New York Jews Won't Stop Winning Nobel Prizes." *Times of Israel*, 9 Dec. 2012.
http://www.timesofisrael.com/new-york-jews-wont-stop-winning-nobel-prizes/.
Accessed Apr-May 2017.

Senor, Dan and Saul Singer. *Start-up Nation: The Story of Israel's Economic Miracle*. McClelland & Stewart, 2011.

Shapiro, Gary. "Can China Eclipse the U.S. on Innovation?" *Forbes*, 11 Jul. 2012.
https://www.forbes.com/sites/garyshapiro/2012/07/11/can-china-eclipse-the-u-s-on-innovation/#7f04416673c2.
Accessed Apr-May 2017.

Shaw, Kira. "Travel Broadens the Mind, But Can It Alter the Brain?" *The Guardian*, 18 Jan. 2016.
https://www.theguardian.com/education/2016/jan/18/travel-broadens-the-mind-but-can-it-alter-the-brain.
Accessed Apr-May 2017.

Tal, Alon. "To Make a Desert Bloom: The Israeli Agricultural Adventure and the Quest for Sustainability." *Aytzim Ecological Judaism*, n.d.
http://aytzim.org/GZA-AlonTal-The_Israeli_Agricultural_Adventure.pdf.
Accessed Apr-May 2017.

"The Will." *Nobelprize.org*, 27 Jun. 2017.
http://www.nobelprize.org/alfred_nobel/will/.
Accessed Apr-May 2017.

"The Wise Rabbi and the Wise Fish (A Jewish Tale)." *UExpress.com*, 31 Dec. 2000.
http://www.uexpress.com/tell-me-a-story/2000/12/31/the-wise-rabbi-and-the-wise
Accessed Apr-May 2017.

Tse, Edward. "The Rise of Entrepreneurship in China." *Forbes*, 5 Apr. 2016. https://www.forbes.com/sites/tseedward/2016/04/05/the-rise-of-entrepreneurship-in-china/#462701d03efc.
Accessed Apr-May 2017.

Tse, Edward. "What Drives China's Innovation?" *Forbes*, 8 Mar. 2016. https://www.forbes.com/sites/tseedward/2016/03/08/what-drives-chinas-innovation-2/#65f5b2fd4dde.
Accessed Apr-May 2017.

Tyre, Peg. "Is Coding the New Second Language?" *Smithsonian.com*, 23 May 2013. http://www.smithsonianmag.com/innovation/is-coding-the-new-second-language-81708064/.
Accessed Apr-May 2017.

Wallace, Jennifer and Lisa Heffernan. "Advice College Admissions Officers Give Their Own Kids." *The New York Times*, 17 Mar. 2016. https://well.blogs.nytimes.com/2016/03/17/advice-college-admissions-officers-give-their-own-kids/?_r=1.
Accessed Apr-May 2017.

Westly, Erica. "The Bilingual Advantage: Second Language Increases Cognitive Ability." *Scientific American*, Jul. 2011. https://www.scientificamerican.com/article/the-bilingual-advantage/.
Accessed Apr-May 2017.

Yin, David. "Out of Israel, Into the World." Forbes Asia. Dec 2013. https://www.forbes.com/sites/davidyin/2013/12/19/out-of-israel-into-the-world/#28677e3c367d .
Accessed Dec 2017.

www.ingramcontent.com/pod-product-compliance
Lightning Source LLC
Chambersburg PA
CBHW031444040426
42444CB00007B/963